Advanced Maths Essentials
Core 1 for Edexcel

Welcome to Advanced Maths Essentials: Core 1 for Edexcel. This book will help you to improve your examination performance by focusing on all the essential maths skills you will need in your Edexcel Core 1 examination. It has been divided by chapter into the main topics that need to be studied. Each chapter has then been divided by sub-headings, and the description below each sub-heading gives the Edexcel specification for that aspect of the topic.

The book contains scores of worked examples, each with clearly set-out steps to help solve the problem. You can then apply the steps to solve the Skills Check questions in the book and past exam questions at the end of each chapter. If you feel you need extra practice on any topic, you can try the Skills Check Extra exercises on the accompanying CD-ROM. At the back of this book there is a sample exam-style paper to help you test yourself before the big day.

Some of the questions in the book have a ⊚ symbol next to them. These questions have a PowerPoint® solution (on the CD-ROM) that guides you through suggested steps in solving the problem and setting out your answer clearly.

Using the CD-ROM

To use the accompanying CD-ROM simply put the disc in your CD-ROM drive, and the menu should appear automatically. If it doesn't automatically run on your PC:

1. Select the My Computer icon on your desktop.
2. Select the CD-ROM drive icon.
3. Select Open.
4. Select core_1_edexcel.exe.

If you don't have PowerPoint® on your computer you can download PowerPoint 2003 Viewer®. This will allow you to view and print the presentations. Download the viewer from http://www.microsoft.com

Pearson Education Limited
Edinburgh Gate
Harlow
Essex
CM20 2JE
England
www.longman.co.uk

© Pearson Education Limited 2005

The rights of Janet Crawshaw, Keith Gordon and Karim Hirani to be identified as the authors of this Work have been asserted by Janet Crawshaw, Keith Gordon and Karim Hirani in accordance with the Copyright, Designs and Patents Act, 1988.

First published 2005
ISBN   0 582 83663 8

Design by Ken Vail Graphic Design

Cover design by Raven Design

Typeset by Tech-Set, Gateshead

Printed in the U.K. by Scotprint, Haddington

The publisher's policy is to use paper manufactured from sustainable forests.

The Publisher wishes to draw attention to the Single-User Licence Agreement situated at the back of the book. Please read this agreement carefully before installing and using the CD-ROM.

We are grateful for permission from London Qualifications Limited trading as Edexcel to reproduce past exam questions. All such questions have a reference in the margin. London Qualifications Limited trading as Edexcel can accept no responsibility whatsoever for accuracy of any solutions or answers to these questions.

Every effort has been made to ensure that the structure and level of sample question papers matches the current specification requirements and that solutions are accurate. However, the publisher can accept no responsibility whatsoever for accuracy of any solutions or answers to these questions. Any such solutions or answers may not necessarily constitute all possible solutions.

1 Algebra and functions

1.1 Indices

Laws of indices for all rational exponents.

The laws or rules of indices allow you to simplify terms that are written in **index form**, a^m, where m is rational.

a is the **base**, where $a \neq 0$.

m is the **index**, also known as the **power** or **exponent**.

These rules apply to terms in index form:

> **Note:**
> The index m can be a positive or negative integer, or a fraction, or zero.

> **Rule 1** To **multiply** the terms, **add** the indices:
> $$a^m \times a^n = a^{m+n}$$
>
> **Rule 2** To **divide** the terms, **subtract** the indices:
> $$a^m \div a^n = a^{m-n}$$
>
> **Rule 3** To **raise to a power**, **multiply** the indices:
> $$(a^m)^n = a^{mn}$$

> **Tip:**
> $a^m \div a^n$ is also written $\dfrac{a^m}{a^n}$.

Example 1.1 Simplify   **a** $2x^5 \times 7x^6$ **b** $10x^5z^3 \div 2x^3z$ **c** $(2p^2)^5$

Step 1: Gather like terms.

Step 2: Simplify using the index laws.

a $\begin{aligned} 2x^5 \times 7x^6 &= 2 \times 7 \times x^5 \times x^6 \\ &= 14 \times x^{5+6} \\ &= 14x^{11} \end{aligned}$ (*Rule 1*)

> **Tip:**
> Multiply or divide the numbers, then deal with the expressions in index form.

b $\begin{aligned} 10x^5z^3 \div 2x^3z &= \frac{10x^5z^3}{2x^3z} \\ &= 5x^{5-3}z^{3-1} \\ &= 5x^2z^2 \end{aligned}$ (*Rule 2*)

> **Tip:**
> When no index is written, this means the power is 1, so $z = z^1$.

c $(2p^2)^5 = 2^5(p^2)^5 = 32p^{10}$ (*Rule 3*)

> **Tip:**
> $(ab)^n = a^n b^n$, so remember to raise 2 to the power 5 here.

Example 1.2 **a** Write each of these expressions as a power of 2:

    i 8^4 **ii** 4^{x+1}

b Hence solve the equation $8^4 = 4^{x+1}$.

Step 1: Write each term in index form with the same base.

a **i** $8^4 = (2^3)^4 = 2^{12}$ (*Rule 3*)

    ii $4^{x+1} = (2^2)^{x+1} = 2^{2(x+1)} = 2^{2x+2}$ (*Rule 3*)

Step 2: Simplify using the index laws.

b $8^4 = 4^{x+1} \Rightarrow 2^{12} = 2^{2x+2}$

Step 3: Equate the indices and solve.

Equating indices gives $12 = 2x + 2$

$$x = 5$$

The zero index, a^0

You know that $\quad a^n \div a^n = 1$

But, by Rule 3, $\quad a^n \div a^n = a^{n-n} = a^0$

$\Rightarrow \qquad\qquad\qquad a^0 = 1$ **Rule 4**

> **Note:**
> a cannot be zero; 0^0 is undefined.

Negative index, a^{-n}

You know that $a^n \times a^{-n} = a^0 = 1$

Divide both sides by a^n $\quad a^{-n} = \dfrac{1}{a^n}$ **Rule 5a**

Note:
$5^{-2} = \dfrac{1}{5^2} = \dfrac{1}{25}$

This format is useful when working in fractions:

$\left(\dfrac{a}{b}\right)^{-n} = \left(\dfrac{b}{a}\right)^{n}$ **Rule 5b**

Note:
$\left(\frac{2}{3}\right)^{-2} = \left(\frac{3}{2}\right)^2 = \frac{9}{4} = 2\frac{1}{4}$

Fractional indices

$a^{\frac{1}{n}} = \sqrt[n]{a}$ **Rule 6a**

For example, $a^{\frac{1}{3}} = \sqrt[3]{a}$.

$a^{\frac{m}{n}} = (\sqrt[n]{a})^m = \sqrt[n]{a^m}$ **Rule 6b**

For example, $a^{\frac{2}{3}} = (\sqrt[3]{a})^2 = \sqrt[3]{(a^2)}$.

To calculate $64^{\frac{2}{3}}$ you could find $(\sqrt[3]{64})^2 = 4^2 = 16$.

Alternatively, you could find $\sqrt[3]{(64^2)} = \sqrt[3]{4096} = 16$.

Tip:
Choose the format that is more convenient, depending on your numbers.

Example 1.3 Evaluate, without using a calculator,

a $3^4 \div 3^7$ **b** $\left(\frac{3}{4}\right)^{-1}$ **c** $4^{\frac{1}{2}}$ **d** $8^{-\frac{1}{3}}$ **e** $\left(\frac{1}{125}\right)^{-\frac{2}{3}}$

Step 1: Calculate using the index laws.

a $3^4 \div 3^7 = 3^{-3} = \dfrac{1}{3^3} = \dfrac{1}{27}$ *(Rules 2 & 5a)*

b $\left(\frac{3}{4}\right)^{-1} = \left(\frac{4}{3}\right)^1 = \frac{4}{3}$ *(Rule 5b)*

c $4^{\frac{1}{2}} = \sqrt{4} = 2$ *(Rule 6a)*

d $8^{-\frac{1}{3}} = \dfrac{1}{8^{\frac{1}{3}}} = \dfrac{1}{\sqrt[3]{8}} = \dfrac{1}{2}$ *(Rules 5a & 6a)*

e $\left(\frac{1}{125}\right)^{-\frac{2}{3}} = 125^{\frac{2}{3}} = (\sqrt[3]{125})^2 = 5^2 = 25$ *(Rules 5b & 6b)*

Example 1.4 Write $\frac{1}{3}9^x 27^{2x-1}$ as a power of 3.

Step 1: Write each term in index form with the same base.

$\frac{1}{3}9^x 27^{2x-1} = 3^{-1}(3^2)^x(3^3)^{2x-1}$

$= 3^{-1} 3^{2x} 3^{6x-3}$

Step 2: Simplify using the index laws.

$= 3^{8x-4}$

Example 1.5 Given that $f(x) = 6x^3 + x$ and $g(x) = \sqrt{x}$, express, in index form,

a $f(x) \times g(x)$ **b** $f(x) \div g(x)$.

Tip:
$\sqrt{x} = x^{\frac{1}{2}}$

Step 1: Expand the brackets.

a $f(x) \times g(x) = (6x^3 + x) \times \sqrt{x}$

$= 6x^3 \times x^{\frac{1}{2}} + x^1 \times x^{\frac{1}{2}}$

Step 2: Simplify using the index laws.

$= 6x^{\frac{7}{2}} + x^{\frac{3}{2}}$ *(Rule 1)*

Tip:
When the index is a mixed number, write it as a top-heavy fraction.

Step 1: Divide each term of the numerator by the denominator.

Step 2: Simplify using the index laws.

b $\mathrm{f}(x) \div \mathrm{g}(x) = \dfrac{6x^3 + x}{\sqrt{x}}$

$$= \dfrac{6x^3}{x^{\frac{1}{2}}} + \dfrac{x}{x^{\frac{1}{2}}}$$

$$= 6x^{\frac{5}{2}} + x^{\frac{1}{2}} \qquad \qquad \textit{(Rule 2)}$$

Note:
These techniques are useful when differentiating and integrating.

SKILLS CHECK **1A: Indices**

1 Simplify the following:

    a $2x^3y^5 \times 3xy^{-1}$ **b** $(2a^2)^4$ **c** $14pq^7 \div 2p^2q^5$

2 Evaluate, without using a calculator:

    a $\dfrac{1}{2^{-1}}$ **b** 3^{-2} **c** $27^{-\frac{1}{3}}$

    d $16^{\frac{3}{2}}$ **e** $0.25^{-\frac{1}{2}}$

 3 Simplify $3a^2b^{-2} \times 4a^3 \sqrt{b}$.

4 Write as a single power of x:

    a $x^2\sqrt{x}$ **b** $\dfrac{\sqrt{x}(\sqrt{x})^3}{x^3}$ **c** $\dfrac{\sqrt{x}(\sqrt{x})^3}{x^{-3}}$

5 Write $\dfrac{p^{\frac{1}{6}}p^{\frac{2}{3}}}{\sqrt{p}}$ in the form p^k where k is a number to be found.

 6 a Write each of the following as a power of 2: **i** 4^x **ii** 8^{x-1}

    b Express $4^x 8^{x-1}$ as a single power of 2.

    c Solve the equation $4^x = 8^{x-1}$.

7 a Express 9^{2x} as a power of 3.

    b Solve $3^{x-1} = 9^{2x}$.

    c Solve $9^{2x} = \dfrac{1}{3^3}$.

8 a Given that $8^{2x-1} = 4^y$, form an equation in the form $y = ax + b$, where a and b are rational numbers to be found.

    b Given also that $9^{x+1} = \dfrac{81^{y-1}}{27}$, form another equation relating x and y.

    c Hence find the values of x and y.

9 Given that $7^{x+6} = 49^{2x}$, find x.

10 a Express 36^{2p} as a power of 6.

    b Express 216^{q-1} as a power of 6.

    c Given that $36^{2p} = 216^{q-1}$, form a linear equation in p and q.

    d Given also that $p = 3q$, find the values of p and q.

SKILLS CHECK **1A EXTRA is on the CD**

Use and manipulation of surds.

Some square roots cannot be written as a fraction or as a terminating or recurring decimal. They are irrational and can be written as **surds**.

Examples of surds are $\sqrt{2}, \sqrt{3}, \sqrt{5}$.

Like surds can be combined by **adding** or **subtracting**:

$$5\sqrt{3} + 4\sqrt{3} + 6\sqrt{5} - 2\sqrt{5} = 9\sqrt{3} + 4\sqrt{5}$$

When **multiplying**, remember that $\sqrt{p} \times \sqrt{q} = \sqrt{p \times q}$.

$$\sqrt{2} \times \sqrt{3} = \sqrt{2 \times 3} = \sqrt{6}$$

When **squaring**, use the special case $\sqrt{p} \times \sqrt{p} = \sqrt{p^2} = p$.

$$(3\sqrt{5})^2 = 3\sqrt{5} \times 3\sqrt{5} = 9 \times \sqrt{5 \times 5} = 9 \times 5 = 45$$

To **simplify** a surd such as $\sqrt{12}$ or $\sqrt{180}$, write it in the form $k\sqrt{a}$, where a does not have any factors that are square numbers.

$$\sqrt{12} = \sqrt{2 \times 2 \times 3} = \sqrt{2^2 \times 3} = \sqrt{2^2} \times \sqrt{3} = 2 \times \sqrt{3} = 2\sqrt{3}$$

$$\sqrt{180} = \sqrt{2 \times 2 \times 3 \times 3 \times 5} = \sqrt{2^2 \times 3^2 \times 5} = 2 \times 3 \times \sqrt{5} = 6\sqrt{5}$$

Example 1.6   **a**   Write $\sqrt{8}$ in the form $k\sqrt{2}$, where k is an integer.

b   Hence simplify $6\sqrt{2} + 5\sqrt{8}$.

Step 1: Simplify the surd.   **a** $\sqrt{8} = \sqrt{4 \times 2} = \sqrt{4} \times \sqrt{2} = 2\sqrt{2}$

Step 2: Add like surds.   **b** $6\sqrt{2} + 5\sqrt{8} = 6\sqrt{2} + 5 \times 2\sqrt{2}$
$$= 6\sqrt{2} + 10\sqrt{2}$$
$$= 16\sqrt{2}$$

Example 1.7   Expand and simplify $(2 - \sqrt{3})(5 + \sqrt{3})$.

Step 1: Expand the brackets. $(2 - \sqrt{3})(5 + \sqrt{3}) = 10 + 2\sqrt{3} - 5\sqrt{3} - (\sqrt{3})^2$
Step 2: Simplify each term. $= 10 - 3\sqrt{3} - 3$
Step 3: Add like terms. $= 7 - 3\sqrt{3}$

If you recognise the **difference of two squares** you can multiply quickly, using $(a + b)(a - b) = a^2 - b^2$, for example

$$(3 + \sqrt{2})(3 - \sqrt{2}) = 3^2 - (\sqrt{2})^2 = 9 - 2 = 7$$

To simplify fractions with a surd in the denominator, use a special technique called **rationalising the denominator**. This involves forming an equivalent fraction by multiplying both the numerator and denominator by a quantity that makes the denominator rational.

If the term is in the form $\dfrac{a}{\sqrt{b}}$ then, to rationalise the denominator,

multiply by $\dfrac{\sqrt{b}}{\sqrt{b}}$ $(= 1)$.

Example 1.8 Express $\dfrac{12}{\sqrt{3}}$ in the form $k\sqrt{3}$, where k is an integer.

Step 1: Rationalise the denominator.
$$\frac{12}{\sqrt{3}} = \frac{12}{\sqrt{3}} \times \frac{\sqrt{3}}{\sqrt{3}}$$

Step 2: Simplify if possible.
$$= \frac{12\sqrt{3}}{3} = 4\sqrt{3}$$

Tip:
Multiplying the denominator by $\sqrt{3}$ makes it rational as $\sqrt{3} \times \sqrt{3} = 3$.

If the term is in the form $\dfrac{a}{b + \sqrt{c}}$ then, to rationalise the denominator,

multiply by $\dfrac{b - \sqrt{c}}{b - \sqrt{c}}$ $(= 1)$.

If the term is in the form $\dfrac{a}{\sqrt{b} + c}$ then, to rationalise the denominator,

multiply by $\dfrac{\sqrt{b} - c}{\sqrt{b} - c}$ $(= 1)$.

Example 1.9 Simplify $\dfrac{2\sqrt{3} + 1}{\sqrt{3} - 1}$.

Step 1: Rationalise the denominator.
$$\frac{2\sqrt{3} + 1}{\sqrt{3} - 1} = \frac{(2\sqrt{3} + 1)}{(\sqrt{3} - 1)} \times \frac{(\sqrt{3} + 1)}{(\sqrt{3} + 1)}$$

Step 2: Simplify.
$$= \frac{6 + 3\sqrt{3} + 1}{3 - 1^2}$$

$$= \frac{7 + 3\sqrt{3}}{2}$$

Tip:
Use the difference of two squares to make the denominator rational.

Example 1.10 A rectangle $ABCD$ has area $20\sqrt{2}$ cm² and length $AB = 4\sqrt{5}$ cm.

Giving your answers in simplified surd form, find **a** the length of BC
b the length of DB.

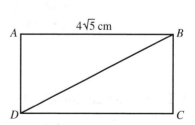

Step 1: Use the formula for the area of a rectangle to find the length.

a
$$\text{Area} = AB \times BC$$
$$\Rightarrow 20\sqrt{2} = 4\sqrt{5} \times BC$$
$$BC = \frac{20\sqrt{2}}{4\sqrt{5}}$$

Step 2: Rationalise the denominator.
$$= \frac{20\sqrt{2} \times \sqrt{5}}{4\sqrt{5} \times \sqrt{5}}$$

Step 3: Simplify.
$$= \frac{20\sqrt{10}}{4 \times 5}$$

$$= \sqrt{10}$$

Length $BC = \sqrt{10}$ cm

Tip:
Remember to simplify any surds where possible.

Step 1: Use Pythagoras' theorem in triangle ADB.

b By Pythagoras' theorem

$$DB^2 = AD^2 + AB^2$$
$$= (\sqrt{10})^2 + (4\sqrt{5})^2$$
$$= 10 + 16\,(5)$$
$$= 90$$
$$DB = \sqrt{90}$$

Step 2: Simplify.

$$= \sqrt{9 \times 10}$$
$$= 3\sqrt{10}$$

Length $DB = 3\sqrt{10}$ cm

Tip:
$$(4\sqrt{5})^2 = 4\sqrt{5} \times 4\sqrt{5}$$
$$= 16 \times (\sqrt{5})^2$$

For further examples using surds, see Sections 1.7 (Quadratic equations) and 1.9 (Inequalities).

SKILLS CHECK 1B: Surds

1 Simplify **a** $\sqrt{50}$ **b** $\sqrt{32}$ **c** $\sqrt{98}$ **d** $\sqrt{12} + 5\sqrt{3}$ **e** $3\sqrt{7} - \sqrt{5} + 4\sqrt{7} - 3\sqrt{5}$.

2 Simplify the following, where $p = \sqrt{3}$, $q = \sqrt{2}$, $r = \sqrt{12}$, $s = \sqrt{18}$.

a $3r + 5p$ **b** $s - q$ **c** $(p - q)^2$ **d** $(5q)^2$

e q^3 **f** $\dfrac{s}{q}$ **g** $\dfrac{1}{r - p}$ 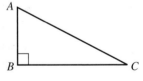 **h** $\dfrac{q}{q + p}$

3 Simplify **a** $(\sqrt{11} + 1)^2$ **b** $(\sqrt{11} + 1)(\sqrt{11} - 1)$ **c** $\dfrac{\sqrt{11} + 1}{\sqrt{11} - 1}$.

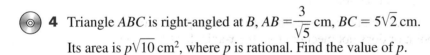 **4** Triangle ABC is right-angled at B, $AB = \dfrac{3}{\sqrt{5}}$ cm, $BC = 5\sqrt{2}$ cm. Its area is $p\sqrt{10}$ cm², where p is rational. Find the value of p.

5 a Express $\sqrt{48}$ and $\dfrac{6}{\sqrt{3}}$ in the form $k\sqrt{3}$, where k is an integer.

b Hence write $\sqrt{48} + \dfrac{6}{\sqrt{3}}$ in the form $p\sqrt{3}$, where p is an integer.

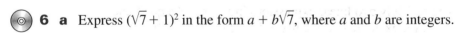 **6 a** Express $(\sqrt{7} + 1)^2$ in the form $a + b\sqrt{7}$, where a and b are integers.

b Express $\dfrac{\sqrt{7} + 1}{\sqrt{7} - 1}$ in the form $c + d\sqrt{7}$, where c and d are rational numbers.

7 a Given that $(3 + \sqrt{5})(4 - \sqrt{5}) = p + q\sqrt{5}$, where p and q are integers, find p and q.

b Given that $\dfrac{3 + \sqrt{5}}{4 + \sqrt{5}} = r + s\sqrt{5}$, where r and s are rational numbers, find r and s.

SKILLS CHECK 1B EXTRA is on the CD

1.3 Quadratic functions

Quadratic functions and their graphs.

The function $f(x) = ax^2 + bx + c$, where $a \neq 0$, is a **quadratic** function.

Note:
If a is zero, there is no x^2 term.

The graph of $y = ax^2 + bx + c$ is called a **parabola**. It is a symmetrical curve with one turning point, called the **vertex**.

If $a > 0$, the turning point is a **minimum** turning point.

If $a < 0$, the turning point is a **maximum** turning point.

Note:
See Section 1.11 for more on finding the turning point and drawing parabolas.

The diagram shows $y = 2x^2 - 4x + 5$.

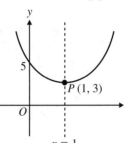

There is a minimum turning point at the vertex $P(1, 3)$.

The minimum value of y is 3.

The axis of symmetry is the line $x = 1$.

1.4 The discriminant

The discriminant of a quadratic function.

Given a quadratic function, $f(x) = ax^2 + bx + c$, the **discriminant** is the expression $b^2 - 4ac$.

It can be used to find

- the number of solutions of the quadratic equation $ax^2 + bx + c = 0$ (Section 1.7)

- whether a quadratic curve crosses the x-axis at two points, touches it at one point or does not meet or cross the x-axis (Section 1.11)

- whether a line and a curve, or two curves, intersect in two places, meet at a point or do not intersect (Section 1.12).

1.5 Factorising quadratic polynomials

When you multiply two linear expressions in x, you get a quadratic polynomial, for example $(3x - 2)(4x + 5) = 12x^2 + 7x - 10$.

To **factorise** a quadratic polynomial, reverse the process and express it as the product of two linear functions.

Note:
Not all quadratic expressions can be factorised.

Some factorising can be done by guesswork, especially when the coefficient of x^2 is 1, for example:

$$x^2 + 8x = x(x + 8) \qquad \text{(common factor)}$$
$$x^2 - 9 = (x - 3)(x + 3) \quad \text{(difference of two squares)}$$
$$x^2 + 8x + 7 = (x + 1)(x + 7)$$
$$x^2 + 8x - 9 = (x - 1)(x + 9)$$

Tip:
Always look for a common factor first.

When guesswork becomes time-consuming, it may be quicker to use a method, as in the following example.

Example 1.11 Factorise $12x^2 + 7x - 10$.

Step 1: Compare with $ax^2 + bx + c$ and find ac.

$a = 12$, $b = 7$, $c = -10$

Step 2: Find factors of ac that add to give b.

$ac = -120$

Factors of -120	Sum
120×-1	119
60×-2	58
20×-6	14
15×-8	7

Tip:

Since ac is negative and b is positive, the signs must be different and the 'larger' number must be positive.

Tip:

Keep practising and you will find that your guesses for the two numbers will become more efficient.

The factors of -120 that add to give 7 are 15 and -8.

Step 3: Rewrite the bx term using the factors found in Step 2.

Replace $7x$ with $15x - 8x$ (or with $-8x + 15x$):

$$12x^2 + 7x - 10 = 12x^2 + 15x - 8x - 10$$

Group the terms in pairs, taking care with signs:

$$12x^2 + 7x - 10 = (12x^2 + 15x) - (8x + 10)$$

Tip:

Remember, if you put a bracket immediately after a minus, you must change the sign in the bracket.

Step 4: Factorise the first two and last two terms. Then take out the common bracket.

Take out any common factors from each of the pairs:

$$12x^2 + 7x - 10 = 3x(4x + 5) - 2(4x + 5)$$

Finally take out a common expression:

$$12x^2 + 7x - 10 = (4x + 5)(3x - 2)$$

Tip:

The expressions in these brackets must be the same.

This method also works when the coefficient of x^2 is negative, as in the following example.

Example 1.12 Factorise $f(x) = -9x^2 + 18x - 5$.

Step 1: Compare with $ax^2 + bx + c$ and find ac.

$a = -9$, $b = 18$, $c = -5$

$ac = -9 \times -5 = 45$

Step 2: Find factors of ac that add to give b.

Factors of 45 that add to give 18 are 15 and 3.

$$-9x^2 + 18x - 5 = -9x^2 + 15x + 3x - 5$$

Step 3: Rewrite the bx term using the factors found in Step 2.

$$= -3x(3x - 5) + 1(3x - 5)$$

$$= (3x - 5)(1 - 3x)$$

Step 4: Factorise the first two and last two terms. Then take out the common bracket.

Important note:

In the special case when $a = 1$, the quadratic expression can be factorised very easily. In this case, ac is the same as c, so you just find two numbers that multiply to give c, the constant term, and add to give b, the x term. These numbers then go in the brackets.

Consider $x^2 - x - 6$:

Factors of -6 that add to give -1 are -3 and 2.

So $x^2 - x - 6 = (x - 3)(x + 2)$.

1.6 Completing the square

Completing the square.

Remember the pattern when squaring linear functions:

$$(x + p)^2 = x^2 + 2px + p^2 \qquad (x - p)^2 = x^2 - 2px + p^2$$

Now rearrange these:

$$x^2 + 2px = (x + p)^2 - p^2 \qquad x^2 - 2px = (x - p)^2 - p^2$$

For example:

$$x^2 + 10x = (x + 5)^2 - 25 \qquad (p = 5)$$
$$x^2 - 6x = (x - 3)^2 - 9 \qquad (p = 3)$$
$$x^2 + 3x = (x + \tfrac{3}{2})^2 - \tfrac{9}{4} \qquad (p = \tfrac{3}{2})$$

> **Recall:**
> $(x + 5)^2 = x^2 + 10x + 25$
> $(x - 3)^2 = x^2 - 6x + 9$

> **Tip:**
> The number in the bracket is half the coefficient of x. Then subtract the square of this number.

This process is called **completing the square**. It can also be applied to quadratic expressions with a constant term, for example:

$$x^2 + 10x + 30 = (x + 5)^2 - 25 + 30 = (x + 5)^2 + 5$$
$$x^2 + 10x - 20 = (x + 5)^2 - 25 - 20 = (x + 5)^2 - 45$$

When $a \neq 1$:

$$3x^2 + 2x + 1 = 3(x^2 + \tfrac{2}{3}x) + 1$$
$$= 3((x + \tfrac{1}{3})^2 - \tfrac{1}{9}) + 1$$
$$= 3(x + \tfrac{1}{3})^2 - \tfrac{1}{3} + 1$$
$$= 3(x + \tfrac{1}{3})^2 + \tfrac{2}{3}$$

> **Tip:**
> Take out a factor of 3 from the x^2 and x terms first.

When $a < 0$:

$$1 - 10x - x^2 = -(x^2 + 10x - 1)$$
$$= -((x + 5)^2 - 25 - 1)$$
$$= -((x + 5)^2 - 26)$$
$$= 26 - (x + 5)^2$$

> **Tip:**
> Take out a factor of -1 first. Remember to multiply through by it at the end.

Completing the square involves writing the quadratic expression $ax^2 + bx + c$ in the form $A(x + B)^2 + C$. You may prefer to complete the square using this identity, as in Example 1.13.

Example 1.13 Write $3x^2 + 2x + 1$ in the form $A(x + B)^2 + C$.

Step 1: Expand $A(x + B)^2 + C$.

$$A(x + B)^2 + C \equiv A(x + B)(x + B) + C$$
$$\equiv A(x^2 + 2Bx + B^2) + C$$
$$\equiv Ax^2 + 2ABx + AB^2 + C$$

> **Note:**
> The $\equiv$ symbol indicates that the expressions are identically equal, for all values of x.

Step 2: Compare coefficients with the original expression.

So $3x^2 + 2x + 1 \equiv Ax^2 + 2ABx + AB^2 + C$

Coefficient of x^2: $\quad 3 = A$

Coefficient of x: $\quad 2 = 2AB \Rightarrow B = \tfrac{1}{3}$

Step 3: Evaluate A, B and C.

Constant term: $\quad 1 = AB^2 + C \Rightarrow C = 1 - AB^2 = 1 - 3(\tfrac{1}{3})^2 = \tfrac{2}{3}$

Substituting $A = 3$, $B = \tfrac{1}{3}$ and $C = \tfrac{2}{3}$ gives

$$3x^2 + 2x + 1 \equiv 3(x + \tfrac{1}{3})^2 + \tfrac{2}{3}$$

Applications of completing the square

When a quadratic function f(x) is in completed square form it is easy to find its maximum or minimum value and also the turning point and axis of symmetry of the curve $y = $ f(x).

In general, $f(x) = A(x + B)^2 + C$ has a turning point at $(-B, C)$.
This is a minimum if $A > 0$ and a maximum if $A < 0$, for example:

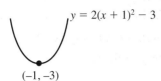

$y = 2(x + 1)^2 - 3$

$(-1, -3)$

$(1, 6)$

$y = -3(x - 1)^2 + 6$

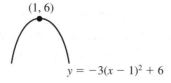

Tip:
If you write $y = -3(x - 1)^2 + 6$ as $y = 6 - 3(x - 1)^2$ it is easier to see that the maximum value is 6.

Example 1.14 It is given that $f(x) = 2x^2 - 4x + 5$.

   a Write $f(x)$ in the form $A(x + B)^2 + C$.

   b Write down the least value of $f(x)$ and state the value of x at which this occurs.

   c Hence state the coordinates of the minimum turning point on the curve $y = f(x)$.

   d Write down the equation of the axis of symmetry of the curve.

Step 1: Expand $A(x + B)^2 + C$.

a $2x^2 - 4x + 5 \equiv A(x + B)^2 + C$
$\equiv Ax^2 + 2ABx + AB^2 + C$

Step 2: Compare coefficients with the original expression, and evaluate A, B and C.

Coefficient of x^2: $2 = A$

Coefficient of x: $-4 = 2AB \Rightarrow B = -1$

Constant term: $5 = AB^2 + C$
$\Rightarrow C = 5 - AB^2 = 5 - 2(-1)^2 = 3$

So $f(x) = 2x^2 - 4x + 5 = 2(x - 1)^2 + 3$.

Step 3: Find the value of x such that $A(x + B)^2 = 0$.

b When $x = 1$, $2(x - 1)^2 = 0$,
so $f(x) = 0 + 3 = 3$.
For *all other values* of x,

Step 4: Substitute into $f(x)$.

$2(x - 1)^2 > 0 \Rightarrow f(x) > 3$.
Since $f(x) \geqslant 3$, the least value of $f(x)$ is 3 and it occurs when $x = 1$.

Tip:
Focus on the squared part of the expression.

Step 5: State $(-B, C)$. **c** The curve $y = 2x^2 - 4x + 5$ has a minimum turning point at $(1, 3)$.

Note:
See Section 1.3 for the sketch of this curve.

Step 6: State the equation of the axis of symmetry.

d The axis of symmetry goes through the turning point so its equation is $x = 1$.

Tip:
See Transformations (Section 1.13).

SKILLS CHECK 1C: Quadratic functions; factorising and completing the square

1 Factorise these expressions.

   a $x^2 + 5x$ **b** $x^2 - 2x + 1$ **c** $a^2 - 16$

   d $x^2 - 5x - 6$ **e** $x^2 + 13x - 30$ **f** $2x^2 - 8x$

2 Factorise these expressions.

   a $2x^2 + 7x + 6$ **b** $5x^2 - 14x - 3$ **c** $3y^2 + 4y - 4$

   d $12 - 4x - 40x^2$ **e** $4x^2 - 25$

3 Write each of these quadratic expressions in the form $A(x + B)^2 + C$.

   a $x^2 + 6x + 8$ **b** $x^2 - 12x - 3$ **c** $x^2 + 5x - 2$

4 Write the quadratic expressions in question **2** in the form $A(x + B)^2 + C$.

5 It is given that $x^2 - 4x + 7 = (x - p)^2 + q$.

   a Find p and q.

   b Hence write down the coordinates of the vertex of the curve $y = x^2 - 4x + 8$.

   c State whether the vertex is a maximum or a minimum turning point.

   d Write down the equation of the axis of symmetry of the curve.

 6 It is given that $f(x) = 4x^2 + 8x + 1$.

   a Write $f(x)$ in the form $A(x + B)^2 + C$.

   b Find the coordinates of the vertex of the curve $y = f(x)$, stating whether it is a maximum or a minimum turning point.

   c Write down the equation of the axis of symmetry of the curve.

 7 **a** Write $f(x) = 14 - 4x - x^2$ in the form $C - (x + B)^2$.

   b Hence state the maximum value of $f(x)$.

   c Write down the coordinates of the vertex of the curve $y = 14 - 4x - x^2$.

SKILLS CHECK **1C EXTRA is on the CD**

1.7 Quadratic equations

Solution of quadratic equations.

A quadratic equation has the form $ax^2 + bx + c = 0$, where $a \neq 0$.

There are several ways of solving quadratic equations. Here are three important techniques.

Solving by factorising

If the quadratic expression can be factorised into the product of two linear functions, solve the quadratic equation using the fact that if $p \times q = 0$, then either $p = 0$ or $q = 0$.

> **Note:**
> Not all quadratic expressions can be factorised.

Example 1.15 **a** Factorise $f(x) = 6x^2 + 11x + 3$.

   b Hence solve $6x^2 + 11x + 3 = 0$.

> **Recall:**
> Section 1.5 for method. Alternatively, factorise by guesswork.

Step 1: Factorise the quadratic expression.

  a $f(x) = 6x^2 + 11x + 3$

        $= 6x^2 + 9x + 2x + 3$

        $= (6x^2 + 9x) + (2x + 3)$

        $= 3x(2x + 3) + 1(2x + 3)$

        $= (2x + 3)(3x + 1)$

Working for factorisation:
$a = 6, b = 11, c = 3$
$ac = 18$
Factors of 18 that add to give 11 are 9 and 2.

  b $6x^2 + 11x + 3 = 0$

Step 2: Use $pq = 0 \Rightarrow$ $p = 0$ or $q = 0$ and solve the resulting linear equations.

$\Rightarrow (2x + 3)(3x + 1) = 0$

$\Rightarrow 2x + 3 = 0 \qquad$ or $\qquad 3x + 1 = 0$

$\qquad\qquad x = -\frac{3}{2} \qquad\qquad\qquad\qquad x = -\frac{1}{3}$

> **Note:**
> You may leave answers as 'top heavy' fractions if you wish.

Example 1.16 Solve $8x^2 + 4x - 2 = -x(2x - 3)$.

Step 1: Rearrange the equation to f(x) = 0.

Step 2: Factorise the quadratic expression.

Step 3: Use $pq = 0 \Rightarrow$ $p = 0$ or $q = 0$ and solve the resulting linear equations.

$$8x^2 + 4x - 2 = -2x^2 + 3x$$
$$10x^2 + x - 2 = 0$$
$$(2x + 1)(5x - 2) = 0$$
$$\Rightarrow 2x + 1 = 0 \qquad \text{or} \qquad 5x - 2 = 0$$
$$x = -\tfrac{1}{2} \qquad\qquad\qquad x = \tfrac{2}{5}$$

Solving by completing the square

Complete the square for the quadratic expression and then rearrange to make x the subject.

Example 1.17 **a** Write $x^2 - 8x + 9$ in the form $(x - p)^2 - q$.

b Hence solve the equation $x^2 - 8x + 9 = 0$, leaving your answers in surd form.

Step 1: Complete the square.

a $x^2 - 8x + 9 = (x - 4)^2 - 16 + 9$
$$= (x - 4)^2 - 7$$

b $$x^2 - 8x + 9 = 0$$
$$\Rightarrow (x - 4)^2 - 7 = 0$$

Step 2: Make x the subject.

(Add 7) $$(x - 4)^2 = 7$$

(Square root both sides) $x - 4 = \pm\sqrt{7}$

(Add 4) $$x = 4 \pm\sqrt{7}$$

The two solutions are $x = 4 + \sqrt{7}$ and $x = 4 - \sqrt{7}$.

Example 1.18 **a** Write $4x^2 + 4x - 15$ in the form $A(x + B)^2 + C$.

b Hence solve $4x^2 + 4x - 15 = 0$.

Step 1: Complete the square.

a Let $4x^2 + 4x - 15 \equiv A(x + B)^2 + C$
$$\equiv Ax^2 + 2ABx + AB^2 + C$$

Coefficient of x^2: $4 = A$

Coefficient of x: $4 = 2AB \Rightarrow B = \tfrac{1}{2}$

Constant term: $-15 = AB^2 + C \Rightarrow C = -15 - 4(\tfrac{1}{2})^2 = -16$

$$4x^2 + 4x - 15 = 4(x + \tfrac{1}{2})^2 - 16$$

Step 2: Make x the subject.

b Rearrange: $4(x + \tfrac{1}{2})^2 = 16$

Divide by 4: $(x + \tfrac{1}{2})^2 = 4$

Take the square root of each side:

$$x + \tfrac{1}{2} = \pm 2$$
$$x = -\tfrac{1}{2} \pm 2$$

Either $x = -\tfrac{1}{2} - 2$ or $x = -\tfrac{1}{2} + 2$
$$= -\tfrac{5}{2} \qquad\qquad = \tfrac{3}{2}$$

Using the quadratic formula

If $ax^2 + bx + c = 0$ then

$$x = \frac{-b \pm \sqrt{b^2 - 4ac}}{2a}$$

You must memorise this formula for your examination.

Note:
The formula is derived by completing the square on $ax^2 + bx + c = 0$.

Example 1.19 Use the quadratic formula to solve $2x^2 + x - 4 = 0$, leaving your answers in surd form.

Step 1: Identify a, b and c. $a = 2, b = 1, c = -4$

Step 2: Substitute into the quadratic formula and evaluate.

$$x = \frac{-b \pm \sqrt{b^2 - 4ac}}{2a}$$

$$= \frac{-1 \pm \sqrt{1^2 - 4(2)(-4)}}{2 \times 2}$$

$$= \frac{-1 \pm \sqrt{33}}{4}$$

$$x = \frac{-1 + \sqrt{33}}{4} \text{ or } \frac{-1 - \sqrt{33}}{4}$$

Note:
These are exact answers. In this case, the values given by a calculator would be approximations.

Number of real roots of a quadratic equation

The term under the square root in the quadratic formula is the **discriminant**, $b^2 - 4ac$. Its value can be used to deduce the number of solutions of the quadratic equation $ax^2 + bx + c = 0$.

Recall:
The discriminant (Section 1.4).

If $b^2 - 4ac > 0$, then $\sqrt{b^2 - 4ac}$ can be calculated and the quadratic formula will give **two different values** for x.
The quadratic equation has **two real distinct roots** (two real solutions).

Tip:
Learn these conditions.

If $b^2 - 4ac = 0$, then $\sqrt{b^2 - 4ac} = 0$ and the quadratic formula will give **only one value** for x.
The quadratic equation has **two real equal roots** (one real solution).

If $b^2 - 4ac < 0$, then there are no real values of $\sqrt{b^2 - 4ac}$.
The quadratic equation has **no real roots** (no real solutions).

Note:
See Section 1.9 (Inequalities) for more on these conditions.

Example 1.20 The equation $2x^2 - 3x + 3k = 0$ has one real solution. Find the value of k.

Step 1: Identify a, b and c. $a = 2, b = -3, c = 3k$

Step 2: Find $b^2 - 4ac$ and use the appropriate condition for the discriminant.

$$b^2 - 4ac = (-3)^2 - 4(2)(3k)$$
$$= 9 - 24k$$

One real solution $\Rightarrow b^2 - 4ac = 0$
$$9 - 24k = 0$$
$$k = \frac{9}{24}$$
$$= \frac{3}{8}$$

1 Solve the following quadratic equations:

a $(x - 2)(x + 3) = 0$ **b** $(1 - 4x)(3 + 2x) = 0$ **c** $4x(x + 5) = 0$

2 Solve the following equations, using the method of factorisation.

a $x^2 + 6x + 5 = 0$ **b** $x^2 - 11x + 24 = 0$ **c** $x^2 - 6x = 0$

d $x^2 - 5x - 6 = 0$ **e** $x^2 - x - 6 = 0$ **f** $x^2 - 36 = 0$

 3 The function f is defined for all x by $f(x) = x^2 + 3x - 5$.

a Express $f(x)$ in the form $(x + P)^2 + Q$.

b Hence, or otherwise, solve the equation $f(x) = 0$, giving your answers in surd form.

4 a Write $2x^2 - 3x - 2$ in the form $A(x + B)^2 + C$.

b Hence solve $2x^2 - 3x - 2 = 0$.

c Check your answers by solving $2x^2 - 3x - 2 = 0$ by the method of factorisation.

5 Use the quadratic formula to solve the equation $5x^2 + 3x - 3 = 0$, leaving your answers in surd form.

 6 a Solve $(2x - 3)^2 = 25$.

b Solve $(2x - 3)^2 = 2x$, expressing your answers in surd form.

7 Find the exact solutions of the quadratic equation $3x^2 + 2x - 4 = 0$.

8 By calculating the discriminant, find the number of real solutions of each of the following quadratic equations:

a $x^2 - 3x + 1 = 0$ **b** $2x^2 - 3x - 1 = 0$ **c** $4x^2 - 4x + 1 = 0$ **d** $5x + x^2 - 3 = 0$

 9 Find the discriminant of $3x^2 - 2x + 5$ and hence show that $3x^2 - 2x + 5 = 0$ has no real solutions.

1.8 Simultaneous equations

Simultaneous equations; analytical solution by substitution.

Linear simultaneous equations

In a pair of simultaneous equations there are two unknowns, for example x and y.

Consider these simultaneous equations.

$2y + 3x = 18$ ①

$5y - x = 11$ ②

To **solve** them you have to find a value of x and a value of y that satisfy *both* equations. The most common methods to use are elimination and substitution.

Elimination method

When the numerical coefficients of one of the unknowns are the same, *eliminate* that unknown either by adding or subtracting the equations.

Step 1: Make the coefficients of one of the unknowns the same then add or subtract to eliminate the unknown.

$$② \times 3 \qquad 15y - 3x = 33 \qquad ③$$
$$2y + 3x = 18 \qquad ①$$
$$③ + ① \qquad \overline{17y \qquad\;\; = 51}$$
$$y = 3$$

Step 2: Substitute the value found into one of the original equations.

Substituting $y = 3$ into equation ①:

$6 + 3x = 18$

$3x = 12 \quad \Rightarrow \quad x = 4$

The solution is $x = 4$, $y = 3$.

Note:
If the terms containing the unknown you want to eliminate have the same sign, subtract. If the signs are different, add.

Note:
The lines $2y + 3x = 18$ and $5y - x = 11$ intersect at $(4, 3)$. See Section 1.12 for more on graphical interpretations.

Substitution method

Use one equation to express one of the unknowns in terms of the other and then *substitute* for it in the other equation.

Step 1: Express one unknown in terms of the other.

Write x in terms of y using equation ②:

$x = 5y - 11$

Step 2: Substitute it into the other equation and solve.

Substituting for x in equation ①:

$2y + 3(5y - 11) = 18$

$2y + 15y - 33 = 18$

$17y = 51 \quad \Rightarrow \quad y = 3$

Step 3: Substitute the value found into one of the original equations.

To find x, proceed as in the elimination method.

Tip:
Although it does not matter whether you write x in terms of y, or y in terms of x, try to avoid expressions with fractions where possible.

One linear and one quadratic equation

Example 1.21 Solve the simultaneous equations.

$2x + y = 6 \qquad ①$

$y = 4 + x - x^2 \qquad ②$

Step 1: Express one unknown in terms of the other using the linear equation.

Step 2: Substitute into the other equation and solve.

From ① $y = 6 - 2x$

Substituting into ② $6 - 2x = 4 + x - x^2$

Rearranging $x^2 - 3x + 2 = 0$

$(x - 1)(x - 2) = 0$

Either $\quad x - 1 = 0 \quad$ or $\quad x - 2 = 0$

$x = 1 \qquad\qquad x = 2$

Recall:
Solving quadratic equations (Section 1.7).

Step 3: Substitute the values found into the equation from step 1.

Substituting for x into $y = 6 - 2x$:

When $x = 1$, $y = 6 - 2 \quad \Rightarrow \quad y = 4$

When $x = 2$, $y = 6 - 4 \quad \Rightarrow \quad y = 2$

Step 4: State the solutions in pairs.

The solutions are $x = 1$, $y = 4$ or $x = 2$, $y = 2$.

Note:
The line and the curve intersect at $(1, 4)$ and $(2, 2)$ (see Section 1.12).

1 Solve these linear simultaneous equations by the method of elimination:

a $8x + 5y = 37$
$2x - 5y = 3$

b $3x + 2y = 13$
$x + y = 5$

c $3a - 2b = 13$
$5a - 7b = 29$

2 Solve these linear simultaneous equations by the method of substitution:

a $y = x + 10$
$3y + 2x = 0$

b $a - 2b = 5$
$2a = 5b + 7$

c $p = 3 - 4q$
$p = q - 12$

 3 **a** Show that the equation $\dfrac{x}{3} + \dfrac{y}{2} = 1$ can be written in the form $2x + 3y = 6$.

b Solve these simultaneous equations for x and y:

$\dfrac{x}{3} + \dfrac{y}{2} = 1$; $x + 3y = 2$

4 Solve for x and y:

a $y + x = 10$
$y = x^2 + 3x - 2$

b $y = 3x$
$xy + x = 2$

 c $y = x + 1$
$x^2 + y^2 = 1$

 5

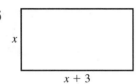

x

$x + 3$

The area of this rectangle is y cm^2
and the perimeter is y cm.
Find the dimensions of the rectangle.

6 A curve has equation $y = ax^2 + bx$, where a and b are integers.

a The point $(3, 21)$ lies on the curve. Show that $7 = 3a + b$.

b The point $(2, 8)$ also lies on the curve. Find another linear equation satisfied by a and b.

c Solve the two equations simultaneously to find the equation of the curve.

1.9 Inequalities

Solution of linear and quadratic inequalities.

Operations on inequalities

If you perform these operations on both sides of an inequality, the inequality is unaffected:

- add a number
- subtract a number
- multiply by a *positive* number
- divide by a *positive* number.

Remember that the inequality is reversed if you do the following:

- multiply by a *negative* number
- divide by a *negative* number.

Never multiply or divide an inequality by *an unknown quantity*, as you do not know whether it is positive or negative!

Tip:
Avoid these two operations if possible, as in Example 1.22b.

Linear inequalities

Example 1.22 Find the values of x for which

    a $3x - 4 > 2$ **b** $4 - x \geqslant 1$ **c** $3 < 2x + 1 < 9$

Step 1: Simplify the inequality.

Step 2: Solve the inequality.

a
$$3x - 4 > 2$$
$(+4)$ $3x > 6$
$(\div 3)$ $x > 2$

b
$$4 - x \geqslant 1$$
$(+ x)$ $4 \geqslant x + 1$
(-1) $3 \geqslant x$
$x \leqslant 3$

c
$$3 < 2x + 1 < 9$$
(-1) $2 < 2x < 8$
$(\div 2)$ $1 < x < 4$

> **Tip:**
> Adding x to both sides avoids dealing with $-x$.

Example 1.23 Solve the inequality $6(a + 3) > 8 - 2(a + 1)$.

Step 1: Expand the brackets.

$6a + 18 > 8 - 2a - 2$
$6a + 18 > 6 - 2a$

Step 2: Simplify and solve the inequality.

$8a + 18 > 6$
$8a > -12$
$a > -\dfrac{3}{2}$

> **Tip:**
> Take care with the signs when expanding.

Quadratic inequalities

Inequalities such as $x^2 < 9$ and $x^2 \geqslant 25$ are the simplest quadratic inequalities to solve.

  a To solve $x^2 < 9$, first consider $x^2 = 9$.

    $x^2 = 9$ when $x = 3$ or $x = -3$.

    For any value between -3 and 3, $x^2 < 9$.

    So the range of values of x for which $x^2 < 9$ is $-3 < x < 3$.

  b To solve $x^2 \geqslant 25$, first consider $x^2 = 25$.

    $x^2 = 25$ when $x = 5$ or $x = -5$.

    If $x \leqslant -5$, $x^2 \geqslant 25$. Also if $x \geqslant 5$, $x^2 \geqslant 25$.

    So the values of x for which $x^2 \geqslant 25$ are $x \leqslant -5$ or $x \geqslant 5$.

This method can be applied to more complicated quadratic inequalities by completing the square.

> **Tip:**
> In **a** the values are 'sandwiched' between -3 and 3, so the solution should be written in one inequality.

> **Tip:**
> In **b** the values are outside the 'sandwich', so give two separate inequalities.

> **Recall:**
> Completing the square (Section 1.6).

Example 1.24 **a** Express $x^2 - 6x + 7$ in the form $(x + a)^2 + b$.

    b Find the range of values of x for which $x^2 - 6x + 7 < 0$.

Step 1: Complete the square.

a $x^2 - 6x + 7 = (x - 3)^2 - 9 + 7$
$= (x - 3)^2 - 2$

b
$$x^2 - 6x + 7 < 0$$

Step 2: Solve the inequality using the completed square format.

$$\Rightarrow (x-3)^2 - 2 < 0$$
$$(x-3)^2 < 2$$
$$-\sqrt{2} < x - 3 < \sqrt{2}$$
$$(+3) \quad 3 - \sqrt{2} < x < 3 + \sqrt{2}$$

So $x^2 - 6x + 7 < 0$ when $3 - \sqrt{2} < x < 3 + \sqrt{2}$.

Another method is to use a sketch. This is illustrated in Example 1.25.

Tip:
Substitute y for $(x-3)$ and use the fact that if $y^2 < 2$, then $-\sqrt{2} < y < \sqrt{2}$.

Recall:
Surds (Section 1.2).

Example 1.25 It is given that $f(x) = x^2 + 2x - 8$.

    a Solve $f(x) = 0$.

    b Sketch $y = f(x)$.

    c Hence solve $x^2 + x - 3 \geqslant 5 - x$.

Note:
Sketching a quadratic curve is described more fully in Section 1.11.

Step 1: Solve the quadratic equation.

a $f(x) = 0$ when $x^2 + 2x - 8 = 0$
$$(x-2)(x+4) = 0$$
$$\Rightarrow \quad x = 2, -4.$$

Step 2: State where the curve crosses the x-axis.

The curve $y = f(x)$ crosses the x-axis at $(2, 0)$ and $(-4, 0)$.

Step 3: Sketch the curve.

b

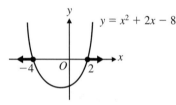

Recall:
Solution of quadratic equations (Section 1.7).

Recall:
The coefficient of x^2 is positive, so the parabola is $\cup$-shaped (Section 1.3).

Tip:
You can use 'filled in' circles to show that x can be 2 or -4.

Step 4: Rearrange the inequality so that there is a zero on the right-hand side.

c
$$x^2 + x - 3 \geqslant 5 - x$$
$$\Rightarrow \quad x^2 + 2x - 8 \geqslant 0$$
$$\Rightarrow \quad f(x) \geqslant 0$$

Step 5: Indicate the possible x-values on the sketch and solve $f(x) \geqslant 0$.

$f(x) \geqslant 0$ when the curve is on or above the x-axis, so from the sketch, $f(x) \geqslant 0$ when $x \leqslant -4$ or $x \geqslant 2$.

Tip:
Write this solution in two separate inequalities.

Application of inequalities to roots of equations

You may need to solve a quadratic inequality in questions about roots of equations, as in the following example.

Example 1.26 Find the values of k for which $2x^2 - kx + 2 = 0$ has no real roots.

Step 1: Identify a, b and c.

Comparing $2x^2 - kx + 2$ with $ax^2 + bx + c$,
$$a = 2, b = -k, c = 2.$$

If the equation has no real roots, then

Recall:
Conditions to be satisfied by the discriminant (Section 1.7).

Step 2: Find $b^2 - 4ac$, and use the appropriate condition for the discriminant.

$$b^2 - 4ac < 0$$
$$\Rightarrow \quad (-k)^2 - 4 \times 2 \times 2 < 0$$
$$k^2 - 16 < 0$$

Step 3: Solve the inequality in k by sketching $y = f(k)$ and finding the values of k when the curve is below the k-axis.

Let $f(k) = k^2 - 16$
$$= (k-4)(k+4)$$
$$f(k) = 0 \text{ when } k = 4 \text{ or } -4$$

The graph of $y = f(k)$ goes through $(-4, 0)$ and $(4, 0)$.

From the sketch, f(k) < 0
when $-4 < k < 4$.

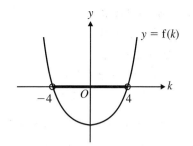

$y = f(k)$

Tip:
You can use 'open' circles to show that k cannot take the values 4 and -4.

So $2x^2 - kx + 2 = 0$ has no real roots when $-4 < k < 4$.

Note:
This solution can be written in one statement.

SKILLS CHECK **1F: Inequalities**

1 Solve these linear inequalities.

  a $4x - 5 > 7$ **b** $3 - 5x < 8$ **c** $2(3x - 1) \geqslant 3(x + 8)$

  d $5y - (4 + y) > 0$ **e** $4 < \dfrac{2x}{3}$ **f** $\dfrac{1}{2}x + \dfrac{3}{4} \leqslant 8$

2 Solve these inequalities:

  a $5 < 2x - 1 < 17$ **b** $-3 \leqslant \dfrac{x}{2} \leqslant 5$

 3 The solution of the inequality $\sqrt{3}\,(x + \sqrt{3}) > 6$ is $x > \sqrt{a}$.
    Find the value of a.

4 Solve these quadratic inequalities.

  a $y^2 > 4$ **b** $x^2 \leqslant 49$ **c** $x^2 \geqslant 5$

  d $2x^2 < 18$ **e** $(x - 1)^2 > 4$ **f** $(x + 2)^2 \leqslant 5$

5 a Express $x^2 + 4x - 5$ in the form $(x + p)^2 + q$, finding the values of the constants p and q.

  b Find the values of x for which $x^2 + 4x \geqslant 5$.

6 Solve these quadratic inequalities.

  a $(x + 4)(x - 3) < 0$ **b** $3(2x + 5)(3x - 2) \geqslant 0$ **c** $(4 - x)(5 + x) \leqslant 0$

  d $p^2 + 7p + 10 < 0$ **e** $2x^2 + x \geqslant 6$

 7 a The quadratic equation $x^2 - 4x - 6 = 0$ has solutions $x = p \pm \sqrt{q}$ where p and q are integers. Find the values of p and q.

  b Sketch $y = x^2 - 4x - 6$.

  c Solve the inequality $x^2 - 4x - 6 < 0$.

8 a Find the values of k for which the equation $3x^2 - kx + 3 = 0$ has no real roots.

  b The equation $kx^2 - 8x + k = 0$ has real roots. Show that $-4 \leqslant k \leqslant 4$.

9 a Find the discriminant of the quadratic expression $2x^2 - kx + 2$.

  b Find the values of k for which $2x^2 - kx + 2 = 0$ has two distinct real roots.

SKILLS CHECK **1F EXTRA is on the CD**

1.10 Algebraic manipulation of polynomials

Algebraic manipulation of polynomials, including expanding brackets, collecting like terms and factorisation.

A **polynomial in x** is an expression with positive integer powers of x, for example $4x^3 + 2x^2 - 7x + 6$. The **degree** of a polynomial is the highest power of x, so this polynomial has degree 3. You may be asked to expand and simplify polynomials.

Note:
The usual convention is to write the polynomial with the highest power first.

Example 1.27

Expand and simplify these polynomials.

a $2x(x^2 + 2x - 1) + (x - 3)(x + 2)$

b $(x - 2)(x + 1)(x - 3)$

c $(x^2 + x - 3)(2x^2 - 4x + 2)$

Step 1: Expand brackets.

Step 2: Collect like terms.

a $2x(x^2 + 2x - 1) + (x - 3)(x + 2)$
$= 2x^3 + 4x^2 - 2x + x^2 + 2x - 3x - 6$
$= 2x^3 + 5x^2 - 3x - 6$

b $(x - 2)(x + 1)(x - 3)$
$= (x^2 - x - 2)(x - 3)$
$= x^3 - 3x^2 - x^2 + 3x - 2x + 6$
$= x^3 - 4x^2 + x + 6$

Tip:
Expand two of the brackets, then multiply by the third.

c $(x^2 + x - 3)(2x^2 - 4x + 2)$
$= 2x^4 - 4x^3 + 2x^2 + 2x^3 - 4x^2 + 2x - 6x^2 + 12x - 6$
$= 2x^4 - 2x^3 - 8x^2 + 14x - 6$

Tip:
Multiplying two brackets, each with three terms, will give nine terms, some of which will combine. Don't try to do this mentally.

Example 1.28

It is given that
$(ax + b)(2x^2 - 3x + 5) \equiv 4x^3 + cx^2 + 7x - 5$.

Find the values of a, b and c.

Note:
This is an identity, true for all values of x.

Step 1: Expand the left-hand side to get the x^3 and constant terms.

Step 2: Equate terms to find a and b.

Consider x^3 and constant terms:
$2ax^3 + \cdots + 5b \equiv 4x^3 + cx^2 + 7x - 5$

Equate x^3 terms: $2a = 4$ $\Rightarrow a = 2$
Equate constants: $5b = -5$ $\Rightarrow b = -1$

Note:
There is only one way to get the x^3 term and this will give the value of a. Similarly, equating the constant terms will give b.

Step 3: Expand to get the x^2 term and solve for c.

Consider x^2 terms:
$\cdots + 2bx^2 - 3ax^2 + \cdots \equiv 4x^3 + cx^2 + 7x - 5$

Equate x^2 terms: $2b - 3a = c$
$-2 - 6 = c \Rightarrow c = -8$

So $a = 2$, $b = -1$ and $c = -8$.

Note:
You could have considered the x terms.

You may be asked to factorise cubic polynomials of the type $ax^3 + bx^2 + cx$. These have a common factor of x, as in the following example.

Example 1.29 Factorise $x^3 - x^2 - 6x$.

Step 1: Look for common factors.

Step 2: Factorise the quadratic expression.

$$x^3 - x^2 - 6x = x(x^2 - x - 6)$$
$$= x(x + 2)(x - 3)$$

Note:
You will learn techniques for factorising $ax^3 + bx^2 + cx + d$ in module C2.

SKILLS CHECK **1G: Algebraic manipulation of polynomials**

1 Expand and simplify the polynomials:

   a $2x(x - 2) - (3x - 1)(x + 5)$ **b** $3x^2(x + 4) + (x^2 - 2x + 1)(x + 4)$

   c $(x + 3)(x - 5)(x - 3)$ **d** $(2x + 1)(x - 3)(x + 4)$

2 Find the values of the letters:

   a $(ax + 3)(x - 4) \equiv 2x^2 - bx - 12$ **b** $(ax^2 + bx + 3)(2x - 1) \equiv 6x^3 - cx^2 + 7x - 3$

   c $(ax + 2)(bx - 1) \equiv 6x^2 + x - c$

3 Factorise fully:

   a $x^3 + 3x^2 - 10x$ **b** $2x^3 + 4x^2 + 2x$ **c** $x^3 - 9x$

   d $6x - 5x^2 - x^3$ **e** $x^3 - 4x^2 - 21x$

SKILLS CHECK **1G EXTRA is on the CD**

1.11 Sketching curves

Graphs of functions; sketching curves defined by simple equations.

To sketch a curve, identify

- what general shape the curve takes
- where the curve crosses the x-axis, by setting $y = 0$
- where the curve crosses the y-axis, by setting $x = 0$.

Note:
You must indicate the intercepts with the axes on the sketch.

Linear functions in *x*

The function $f(x) = ax + b$, where $a \neq 0$, is a **linear** polynomial in x. The graph of $y = ax + b$ is a straight line, with y-intercept b and gradient a.

$$y = ax + b$$

Note:
The highest power of x is 1.

Note:
For more on lines, see Chapter 2: Coordinate geometry in the x, y plane.

Note:
See Section 2.1 for other formats of the equation of a straight line.

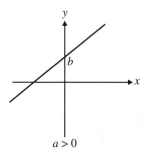

$a > 0$

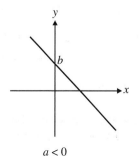

$a < 0$

21

Quadratic functions in *x*

The function $f(x) = ax^2 + bx + c$, where $a \neq 0$, is a **quadratic polynomial** in x and the graph of $y = f(x)$ is a **parabola**.

$a > 0$ $a < 0$

Note:
The highest power of x is 2.

Recall:
Quadratic functions (Section 1.3).

Intercepts with the axes:

$x = 0 \Rightarrow y = c$, so the curve passes through $(0, c)$.

$y = 0 \Rightarrow ax^2 + bx + c = 0$, so the intercepts on the x-axis are found by solving the quadratic equation.

Recall:
Quadratic equations (Section 1.7).

Note that the axis of symmetry of the parabola is a vertical line passing through the midpoint between the intercepts on the x-axis.

Example 1.30 It is given that $f(x) = 2x^2 + 5x - 3$.

     a Sketch $y = f(x)$, labelling the intercepts with the axes.

     b Draw the axis of symmetry on the sketch and state its equation.

Step 1: Decide the general shape.

     a $f(x)$ is a polynomial of degree 2, so the curve is a parabola.
        $a > 0 \Rightarrow$ parabola is $\cup$-shaped.

Recall:
Quadratic functions (Section 1.3).

Step 2: Set $y = 0$ and $x = 0$ to find the axes intercepts.

When $x = 0$, $y = -3$, so the curve goes through $(0, -3)$.

When $y = 0$, $\qquad 2x^2 + 5x - 3 = 0$

$\qquad\qquad\qquad (2x - 1)(x + 3) = 0$

$\qquad\qquad \Rightarrow \quad x = 0.5 \text{ or } x = -3$

Note:
The axis of symmetry and the vertex can also be found by completing the square (Section 1.6).

Step 3: Sketch the curve, marking the intercepts.

$(0.5, 0)$ and $(-3, 0)$ lie on the curve.

Step 4: Draw the axis of symmetry on the sketch and calculate its equation using the x-axis intercepts.

     b Midpoint of x-intercepts $= \dfrac{-3 + 0.5}{2} = -1.25$

        The axis of symmetry is the line $x = -1.25$.

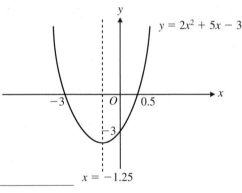

Significance of the discriminant

The discriminant can be used to determine whether or not the curve crosses or touches the x-axis.

If $b^2 - 4ac > 0$, the equation $ax^2 + bx + c = 0$ has two distinct (different), real roots. These give the two x-coordinates of the points where the graph of $y = ax^2 + bx + c$ crosses the x-axis.

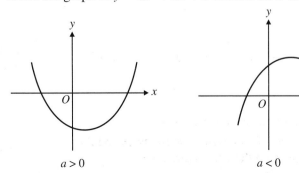

$a > 0$ $a < 0$

If $b^2 - 4ac = 0$, then $ax^2 + bx + c = 0$ has two equal real roots (one real solution). This gives the x-coordinate where the graph of $y = ax^2 + bx + c$ touches the x-axis.

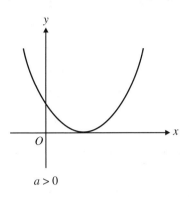

$a > 0$

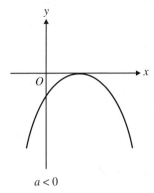

$a < 0$

Note:
The x-axis is a tangent to the curve (see Section 4.4).

If $b^2 - 4ac < 0$, then $ax^2 + bx + c = 0$ has no real roots.

This tells you that the graph of $y = ax^2 + bx + c$ does not cross or touch the x-axis.

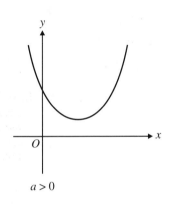

$a > 0$

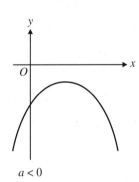

$a < 0$

Example 1.31 Show that the curve $y = 3x^2 - 2x + 5$ lies entirely above the x-axis.

Step 1: Set $x = 0$ and $y = 0$ to attempt to find axes intercepts.

When $x = 0$, $y = 5$, so the curve passes through $(0, 5)$.

When $y = 0$, $3x^2 - 2x + 5 = 0$.
$a = 3$, $b = -2$, $c = 5$

Step 2: Find the discriminant.

$b^2 - 4ac = (-2)^2 - 4 \times 3 \times 5 = 4 - 60 = -56$

Step 3: Use condition on discriminant.

Since $b^2 - 4ac < 0$, the equation $3x^2 - 2x + 5 = 0$ has no real roots and the curve does not cross or touch the x-axis.

Since the curve passes through $(0, 5)$, it must lie entirely above the x-axis.

Note:
You could find the vertex by completing the square (Section 1.6).

Cubic functions

The function $f(x) = ax^3 + bx^2 + cx + d$, where $a \neq 0$, is a **cubic** polynomial in x.

Note:
The highest power of x is 3.

Intercepts with the axes:

$x = 0 \Rightarrow y = d$, so the curve passes through $(0, d)$.

$y = 0 \Rightarrow ax^3 + bx^2 + cx + d = 0$, so to find the intercepts on the x-axis, you would need to solve the cubic equation.

Note:
This is studied in module C2.

In general, the graph of $y = ax^3 + bx^2 + cx + d$ can have the following shapes:

$a > 0$ $a < 0$

Note:
The cubic curve crosses the *x*-axis at least once and no more than three times.

Example 1.32   Sketch the curve $y = (x - 1)(2x + 1)(2 - x)$.

The highest power of x is 3, so the curve is a cubic.

Coefficient of x^3 is -2, so the general shape is one of those described above for $a < 0$.

Note:
A cubic curve may have two, one or no stationary points.

When $y = 0$, $(x - 1)(2x + 1)(2 - x) = 0$

$\Rightarrow$ $x = 1, x = -\frac{1}{2}$ or $x = 2$.

When $x = 0$, $y = (-1) \times 1 \times 2 = -2$.

The curve goes through $(0, -2)$, $(-\frac{1}{2}, 0)$, $(1, 0)$ and $(2, 0)$.

Step 1: Decide the general shape.

Step 2: Set $y = 0$ and $x = 0$ to find the axes intercepts.

Tip:
Multiplying the terms indicated gives the coefficient of x^3:
$(x - 1)(2x + 1)(2 - x)$.

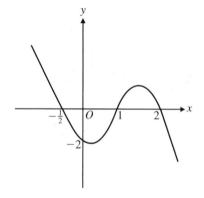

Step 3: Sketch the curve, marking the intercepts.

The reciprocal function

The function $f(x) = \dfrac{k}{x}$, where $x \neq 0$, is called a **reciprocal function** (sometimes referred to as a **hyperbolic function**).

The graph of a reciprocal function, $y = \dfrac{k}{x}$, has no intercepts with the axes: if you substitute $x = 0$, then $y = \dfrac{k}{0}$ is undefined. A similar result is obtained when you substitute $y = 0$ into the equation.

When $k > 0$, if larger and larger positive values of x are substituted, the values of $y = \dfrac{k}{x}$ become smaller and smaller (but remain positive) and tend to 0. This can be written:

as $x \to +\infty, \quad y \to 0+$

Similarly,

as $x \to -\infty, \quad y \to 0-$

Note:
When $k < 0$,
as $x \to +\infty, y \to 0-$
Similarly,
as $x \to -\infty, y \to 0+$

So for large positive and large negative values of x, the curve $y = \dfrac{k}{x}$ approaches the line $y = 0$ (this is the equation for the x-axis).

$y = 0$ is called an **asymptote** to the curve.

When $k > 0$, if smaller and smaller positive values of x are substituted, the values of $y = \dfrac{k}{x}$ become larger and larger (remaining positive), and tend to infinity. This can be written:

$$\text{as } x \to 0+, \quad y \to +\infty$$

Similarly,

$$\text{as } x \to 0-, \quad y \to -\infty$$

So for small positive and small negative values of x, the curve $y = \dfrac{k}{x}$ approaches the line $x = 0$ (this is the equation for the y-axis).
$x = 0$ is an asymptote to the curve.

The sketch of the curve $y = \dfrac{k}{x}$ is as shown.

Note:
The line $y = 0$ is the x-axis.

Note:
When $k < 0$,
as $x \to 0+, y \to -\infty$
Similarly,
as as $x \to 0-, y \to +\infty$

Note:
The line $x = 0$ is the y-axis.

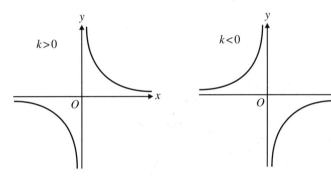

1.12 Geometrical interpretation of algebraic solution of equations

Geometrical interpretation of algebraic solution of equations and use of intersection points of graphs of functions to solve equations.

At a point where two curves, or a line and a curve, intersect or meet, the equation of each must be satisfied.

An algebraic way of finding the point of intersection is to solve the equations simultaneously.

The converse is also true:
If you solve two equations simultaneously you are effectively finding the point of intersection of the curves represented by the equations.

Recall:
Simultaneous equations
(Section 1.8).

Example 1.33　The diagram shows a sketch of $y = x^2 - 9$ and $y = 4x - 12$.

The line and the curve intersect at A and B.

Find the coordinates of A and B.

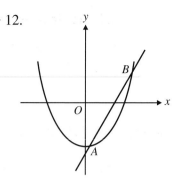

Step 1: Solve the equations simultaneously.

When the line and curve intersect:

$$x^2 - 9 = 4x - 12$$
$$x^2 - 4x + 3 = 0$$
$$(x - 1)(x - 3) = 0$$
$$\Rightarrow \quad x = 1 \text{ or } x = 3$$

Substituting into $y = 4x - 12$:

When $x = 1$, $y = 4 \times 1 - 12 = -8$

When $x = 3$, $y = 4 \times 3 - 12 = 0$

Step 2: Use the solutions to write down the point(s) of intersection.

A is the point $(1, -8)$ and B is the point $(3, 0)$.

SKILLS CHECK　**1H: Sketching curves**

1　Sketch the following, showing the points at which the graph meets the coordinate axes.

   a　$y = 3x + 2$　　　　**b**　$y = 1 - 2x$　　　　**c**　$y = (x - 3)(2x + 1)$

   d　$y = (3 - x)(2 + x)$　　**e**　$y = x^2 + 4x - 5$　　**f**　$y = 7 - 5x + 2x^2$

2　**a**　Factorise $-2x^2 - 7x + 4$.

   b　Sketch the graph of $y = -2x^2 - 7x + 4$, showing the points at which the graph meets the coordinate axes.

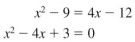 **3**　Sketch the following graphs, showing the vertex and the coordinates of any points at which the graph meets the coordinate axes.

   a　$y = (x - 3)^2$　　　　**b**　$y = (3 - x)^2$

4　It is given that $f(x) = x^2 + 2x - 24$.

   a　Factorise $f(x)$ and hence solve $f(x) = 0$.

   b　Write $f(x)$ in the form $(x + B)^2 + C$.

   c　Using your answers from parts **a** and **b**, sketch $y = f(x)$, stating the coordinates of any points at which the graph meets the coordinate axes.

   d　Write down the equation of the axis of symmetry.

5　Sketch the following cubic graphs, showing the points at which the graph meets the coordinate axes.

   a　$y = (x + 3)(x + 4)(x - 2)$　　**b**　$y = (x + 1)(6 - x)(2 - 3x)$　　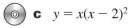 **c**　$y = x(x - 2)^2$

6　Sketch the following curves on the same axes.

   a　$y = \dfrac{12}{x}$　　　　　　　　**b**　$y = -\dfrac{4}{x}$

7 Solve the following pairs of simultaneous equations to find where the line and curve intersect.

a $y = x^2 + 2x - 3$
$y = 4x - 4$

b $y = 2x^2$
$y = 8x - 8$

SKILLS CHECK **1H EXTRA** is on the CD

1.13 Transformations

Knowledge of the effect of simple transformations on the graph of $y = f(x)$ as represented by $y = af(x)$, $y = f(x) + a$, $y = f(x + a)$, $y = f(ax)$.

Translations

$y = f(x) + a$

The transformation $y = f(x) + a$ has the effect of **translating** the graph of $y = f(x)$ by a units in the y-direction.

The vector form of the translation is $\begin{pmatrix} 0 \\ a \end{pmatrix}$.

If $a > 0$, the graph moves up.
If $a < 0$, the graph moves down.

For example:
$y = x^2 + 2$ is a translation
of $y = x^2$ by 2 units up.

$y = x^2 - 1$ is a translation
of $y = x^2$ by 1 unit down.

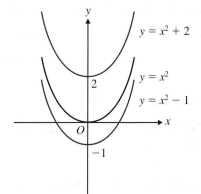

$y = f(x + a)$

The transformation $y = f(x + a)$ has the effect of translating the graph of $y = f(x)$ by $-a$ units in the x-direction.

The vector form of the translation is $\begin{pmatrix} -a \\ 0 \end{pmatrix}$.

If $a > 0$, the graph moves to the left.
If $a < 0$, the graph moves to the right.

For example:
$y = (x - 2)^2$ is a translation
of $y = x^2$ by 2 units to the right.

$y = (x + 1)^2$ is a translation
of $y = x^2$ by 1 unit to the left.

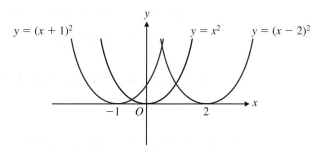

Although you may be examined on the effect of one transformation on a given curve, it is useful to note the effect of performing two translations, one in the x-direction and one in the y-direction.

This is especially useful when sketching quadratic curves when the equation of the curve has been written in 'completed square' form, $y = (x + a)^2 + b$. This represents a translation of $y = x^2$ by $-a$ units in the x-direction and b units in the y-direction.

Note:
The vertex moves from $(0, 0)$ to $(-a, b)$.

This is illustrated in the following example.

Example 1.34 **a** Express $x^2 - 4x - 1$ in the form $(x + a)^2 + b$.

b State the transformation that maps $y = x^2$ onto $y = x^2 - 4x - 1$.

c Hence, or otherwise, sketch the graph of $y = x^2 - 4x - 1$, stating the coordinates of P, the minimum point on the curve.

Step 1: Complete the square. **a** $x^2 - 4x - 1 = (x - 2)^2 - 4 - 1 = (x - 2)^2 - 5$

Step 2: Compare with $(x + a)^2 + b$. **b** $y = x^2 - 4x - 1$ is a translation of $y = x^2$ by 2 units to the right and 5 units down.

Step 3: Sketch the curve and write down the coordinates of the minimum point. **c** The vertex moves from $(0, 0)$ to $(2, -5)$, so the coordinates of P are $(2, -5)$.

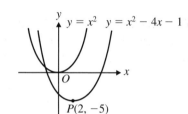

Tip:
The negative of the value inside the bracket gives the x-translation and the constant term gives the y-translation.

Recall:
The minimum point on the curve $y = (x + a)^2 + b$ is $(-a, b)$ (Section 1.6).

Stretches

$y = af(x)$

The transformation $y = af(x)$ has the effect of **stretching** the graph of $y = f(x)$ by a factor of a units in the y-direction. Points on the x-axis are invariant.

Note:
Invariant points do not move under the transformation.

For example, consider $y = x^3$ and $y = 2x^3$.

x	-2	-1	0	1	2
$y = x^3$	-8	-1	0	1	8
$y = 2x^3$	-16	-2	0	2	16

Note:
You can see from the table that, for $y = 2x^3$, all the y-coordinates are multiplied by 2.

The effect is to stretch the curve $y = x^3$ by 2 units in the y-direction.

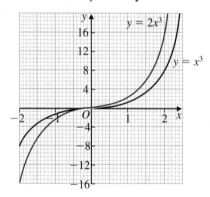

$y = f(ax)$

The transformation $y = f(ax)$ has the effect of stretching the graph of $y = f(x)$ by a factor of $\dfrac{1}{a}$ units in the x-direction. Points on the y-axis are invariant.

When $a > 1$, the graph appears more squashed in the x-direction.
For $0 < a < 1$, the graph appears to be lengthened in the x-direction.

It is important to remember that the correct description in both cases is a *stretch* in the x-direction.

For example, the diagram shows a sketch of $y = f(x)$. A is the point $(-4, 0)$, B is the point $(-1, 5)$ and C is the point $(2, 0)$.

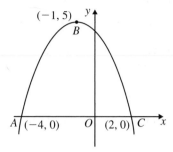

To obtain the graph of $y = f(2x)$, divide all the x-coordinates by 2 while keeping the same y-coordinates.
A moves to $(-2, 0)$, B moves to $(-0.5, 5)$ and C moves to $(1, 0)$.

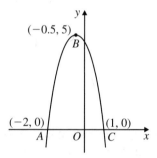

> **Note:**
> This is called a stretch, even though it looks squashed up.

Reflections

Reflections in the x- and y-axes are illustrated below, using the line $y = x + 2$.

$y = -f(x)$

The transformation $y = -f(x)$ is a reflection in the x-axis of $y = f(x)$.

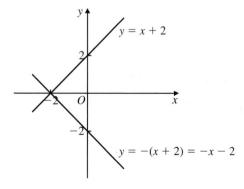

In the graph of $y = -(x + 2)$, all the points in the graph of $y = x + 2$ have been reflected in the x-axis.

$y = f(-x)$

The transformation $y = f(-x)$ is a reflection in the y-axis of $y = f(x)$.

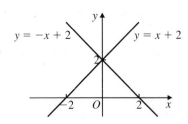

In the graph of $y = -x + 2$, all the points in the graph of $y = x + 2$ have been reflected in the y-axis.

Mixed examples

Example 1.35 The diagram shows a sketch of $y = x(2 - x)$.

The vertex P is the point $(1, 1)$.

By applying appropriate transformations, sketch the following curves:

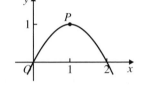

a $y = 3x(2 - x)$ **b** $y = x(2 - x) - 1$.

In each case, describe the transformation and state the coordinates of P', the vertex of the curve.

Step 1: Identify the transformation.

Step 2: Sketch the curve, by applying the transformation.

Step 3: State the coordinates of the vertex.

a

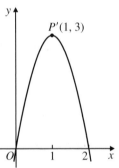

$y = 3x(2 - x)$ is a stretch in the y-direction, factor 3.
The coordinates of P' are $(1, 3)$.

> **Tip:**
> The y-coordinates are multiplied by 3.

b

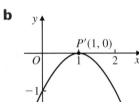

$y = x(2 - x) - 1$ is a translation by 1 unit down in the y-direction, i.e. $\begin{pmatrix} 0 \\ -1 \end{pmatrix}$.
The coordinates of P' are $(1, 0)$.

Example 1.36 The diagram shows the graph of $y = \dfrac{1}{x}$.

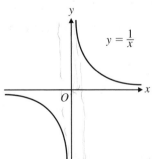

a State the transformation that maps $y = \dfrac{1}{x}$ onto $y = \dfrac{1}{x+3}$.

b Sketch the graph of $y = \dfrac{1}{x+3}$, labelling the asymptotes clearly.

Step 1: Identify the transformation.

a The transformation that maps $y = \dfrac{1}{x}$ onto $y = \dfrac{1}{x+3}$ is a translation by 3 units to the left.

Note:
If $f(x) = \dfrac{1}{x}$, then
$$f(x+3) = \dfrac{1}{x+3}.$$

Step 2: Sketch the curve by applying the transformation.

b

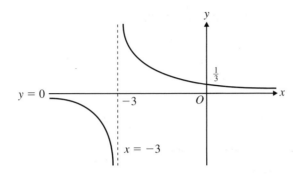

Tip:
When $x = 0$,
$$y = \dfrac{1}{0+3} = \dfrac{1}{3}$$

Example 1.37 The diagram shows a sketch of $y = f(x)$ for $0 \leqslant x \leqslant 3$. For all other values of x, $f(x) = 0$.

A is the point $(2, 1)$ and B is the point $(3, 0)$.

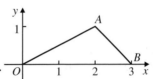

Describe the transformations and draw sketches to show

a $y = f(2x)$ **b** $y = f(\tfrac{1}{3}x)$ **c** $y = f(x-2)$ **d** $y = -f(x)$

In each case, state the new coordinates of A and B.

Tip:
In **a**, all the x-coordinates are halved while maintaining the same y-coordinates.

Step 1: Identify the transformation.

a

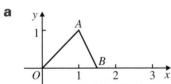

This is a stretch in the x-direction, factor $\tfrac{1}{2}$.
A moves to $(1, 1)$ and B moves to $(1.5, 0)$.

Tip:
In **b**, the factor of stretch $= \dfrac{1}{\frac{1}{3}} = 3$.

Step 2: Draw the sketch by applying the transformation.

Step 3: State the new coordinates.

b This is a stretch in the x-direction, factor 3.

Tip:
In **b**, all the x-coordinates are multiplied by 3 while maintaining the same y-coordinates.

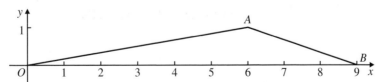

A moves to $(6, 1)$ and B moves to $(9, 0)$.

c This is a translation by 2 units to the right (in the positive x-direction).

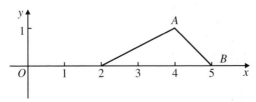

A moves to $(4, 1)$ and B moves to $(5, 0)$.

d This is a reflection in the *x*-axis.

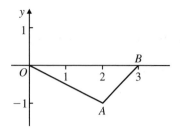

A moves to (2, −1) and *B* stays at (3, 0).

Tip:
Remember to sketch the graph within the given range of *x*-values.

SKILLS CHECK 1I: Transformations

1 For each of the following equations of curves
  i describe the geometrical transformation by which the curve can be obtained from the parabola with equation $y = x^2$,
  ii sketch the curve, stating the coordinates of the vertex and the *y*-intercept.

  a $y = x^2 + 3$ **b** $y = x^2 - 2$ **c** $y = x^2 + 1$

2 Each of the following curves is a translation of $y = x^3$. Sketch the curve and describe the translation.

  a $y = (x + 2)^3$ **b** $y = (x - 1)^3$ **c** $y = (x + 4)^3$

3 Each of the following curves can be written in the form $y = (x - a)^2 + b$. For each curve, state the values of *a* and *b*.

  a

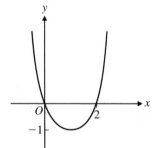

  b

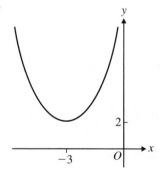

 4 **a** Given that $f(x) = x^2 - 6x + 10$, express $f(x)$ in the form $y = (x + a)^2 + b$.

  b The curve $y = x^2 - 6x + 10$ is a translation of $y = x^2$. Describe the translation and state the minimum point of the curve $y = x^2 - 6x + 10$.

  c By considering the graph of $y = x^2 - 6x + 10$, explain why the equation $x^2 - 6x + 10 = 0$ has no real roots.

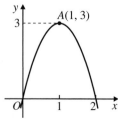

 5 The diagram shows the graph $y = f(x)$ for $0 \leqslant x \leqslant 2$. The vertex *A* is the point (1, 3).

  a Sketch the graph of $y = -f(x)$ for $0 \leqslant x \leqslant 2$ and state the coordinates of the vertex of the curve $y = -f(x)$.

  b **i** Describe fully the transformation that maps $y = f(x)$ onto $y = 2f(x)$
  ii Sketch the graph of $y = 2f(x)$ for $0 \leqslant x \leqslant 2$ and state the coordinates of the vertex of the curve $y = 2f(x)$.

6 a Given that $f(x) = x^2$, state the transformation that maps $y = f(x)$ onto
    i $y = 4f(x)$ **ii** $y = f(2x)$.

   b State the equation of each of the transformed curves and comment.

7 The sketch of $y = f(x)$ for $-2 \leqslant x \leqslant 2$ is drawn below.

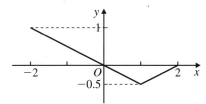

On separate axes, for $-2 \leqslant x \leqslant 2$, sketch the graphs of

   a $y = f(-x)$ **b** $y = -f(x)$ **c** $y = f(x) - 1$.

8 Given that $f(x) = \dfrac{1}{x}$, sketch, on separate graphs, the curves represented by:

   a $y = f(-x)$ **b** $y = \dfrac{1}{3}f(x)$ **c** $y = f(x) - 3$ **d** $y = f(x + 2)$ **e** $y = f(2x)$

On each diagram, also sketch $y = f(x)$ and label the asymptotes of the new graph clearly.

9 By sketching, on the same axes, the graphs of $y = \dfrac{1}{x - 3}$ and $y = \dfrac{1}{x} + 2$, show that there are two

solutions, one negative and one positive, of the equation:

$$\frac{1}{x - 3} = \frac{1}{x} + 2$$

SKILLS CHECK **1I EXTRA** is on the **CD**

Examination practice   Algebra and functions

1 Given that $27^{3x + 1} = 9^y$,

   a obtain an expression for y in the form $y = ax + b$, where a and b are constants,

   b solve the equation $27^{3x + 1} = 9$, giving your answer as an exact fraction.   [Edexcel June 1998]

 2 a Given that $8^y = 4^{2x + 1}$, find y in the form $y = px + q$, where p and q are exact fractions.

   b Solve, giving your answers as exact fractions, the simultaneous equations
       $8^y = 4^{2x + 1}$
       $27^{2y} = 9^{x - 3}$ [Edexcel June 1999]

3 Given that $(2 + \sqrt{7})(4 - \sqrt{7}) = a + b\sqrt{7}$, where a and b are integers,

   a find the value of a and the value of b.

   Given that $\dfrac{2 + \sqrt{7}}{4 + \sqrt{7}} = c + d\sqrt{7}$, where c and d are rational numbers,

   b find the value of c and the value of d.   [Edexcel January 2001]

4 a Find, as surds, the roots of the equation

$$2(x + 1)(x - 4) - (x - 2)^2 = 0.$$

b Hence find the set of values of x for which

$$2(x + 1)(x - 4) - (x - 2)^2 > 0.$$ [Edexcel January 1998]

5 Show that the elimination of x from the simultaneous equations

$$x - 2 = 1,$$
$$3xy - y^2 = 8,$$

produces the equation

$$5y^2 + 3y - 8 = 0.$$

Solve this quadratic equation and hence find the pairs (x, y) for which the simultaneous equations are satisfied. [Edexcel May 1995]

6 Solve the simultaneous equations

$$x + 4y = 10$$
$$x^2 - 2xy - 24y^2 = -250.$$

 7 a Prove, by completing the square, that the roots of the equation $x^2 + 2kx + c = 0$, where k and c are constants, are $-k \pm \sqrt{(k^2 - c)}$.

The equation $x^2 + 2kx + 81 = 0$ has equal roots.

b Find the possible values of k. [Edexcel January 2001]

8 Given that for all values of x,

$$3x^2 + 12x + 5 \equiv p(x + q)^2 + r,$$

a find the values of p, q and r.

b Hence, or otherwise, find the minimum value of $3x^2 + 12x + 5$.

c Solve the equation $3x^2 + 12x + 5 = 0$, giving your answers to one decimal place. [Edexcel January 1996]

9 a Use algebra to solve $(x - 1)(x + 2) = 18$.

b Hence, or otherwise, find the set of values of x for which
$(x - 1)(x + 2) > 18$. [Edexcel June 1998]

10

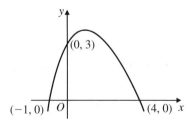

The curve with equation $y = f(x)$ meets the coordinate axes at the points $(-1, 0)$, $(4, 0)$ and $(0, 3)$, as shown in the diagram. Using a separate diagram for each, sketch the curve with equation

a $y = f(x - 1)$, **b** $2y = -f(x)$.

On each sketch, write in the coordinates of the points at which the curve meets the coordinate axes. [Edexcel June 1997]

2 Coordinate geometry in the (x, y) plane

2.1 The equation of a straight line

Equation of a straight line, including the forms $y - y_1 = m(x - x_1)$ and $ax + by + c = 0$.

Midpoint

The midpoint of the line segment joining $A(x_1, y_1)$ and $B(x_2, y_2)$ is given by the formula

$$\text{Midpoint} = \left(\frac{x_1 + x_2}{2}, \frac{y_1 + y_2}{2}\right)$$

Example 2.1 Find the midpoint of AB where A is the point $(7, -3)$ and B is the point $(-2, -5)$.

Step 1: Substitute in the midpoint formula.

Taking $(7, -3)$ as (x_1, y_1) and $(-2, -5)$ as (x_2, y_2):

$$\text{Midpoint} = \left(\frac{x_1 + x_2}{2}, \frac{y_1 + y_2}{2}\right)$$

$$= \left(\frac{7 + (-2)}{2}, \frac{(-3) + (-5)}{2}\right)$$

$$= \left(\frac{5}{2}, -4\right)$$

Distance between two points

The distance AB between two points $A(x_1, y_1)$ and $B(x_2, y_2)$ is given by the formula

$$AB = \sqrt{(x_2 - x_1)^2 + (y_2 - y_1)^2}$$

Example 2.2 Find the distance between A and B where A is the point $(7, -3)$ and B is the point $(-2, -5)$.

Step 1: Substitute into the formula.
Step 2: Work out each bracket, square, add, then square root.

Taking $(7, -3)$ as (x_1, y_1) and $(-2, -5)$ as (x_2, y_2):

$$AB = \sqrt{(-2 - 7)^2 + (-5 - (-3))^2}$$

$$= \sqrt{(-9)^2 + (-2)^2}$$

$$= \sqrt{85}$$

Example 2.3 P is the point $(-6, 1)$ and Q is the point $(8, -1)$. The length PQ is $k\sqrt{2}$. Find the value of k.

Step 1: Substitute into the formula.
Step 2: Work out each bracket, square, add, then square root.

Taking $(-6, 1)$ as (x_1, y_1) and $(8, -1)$ as (x_2, y_2):

$$AB = \sqrt{(8 - (-6))^2 + (-1 - 1)^2}$$

$$= \sqrt{14^2 + (-2)^2}$$

$$= \sqrt{200}$$

$$= \sqrt{100 \times 2}$$

$$= 10\sqrt{2}$$

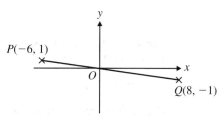

Therefore $k = 10$.

Gradient

The gradient of the line joining $A(x_1, y_1)$ and $B(x_2, y_2)$ is given by the formula

Gradient of $AB = \dfrac{y_2 - y_1}{x_2 - x_1}$

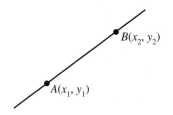

Tip:
Keep the order the same and be careful with minuses.

Example 2.4 Find the gradient of the line AB where A is the point $(7, -3)$ and B is the point $(-2, -5)$.

Step 1: Substitute into the formula.

Gradient of $AB = \dfrac{-5 - (-3)}{-2 - 7} = \dfrac{-2}{-9} = \dfrac{2}{9}$

Tip:
Substitute into the formula before attempting to work anything out.

Equations of lines

There are several ways of writing the equation of a straight line.

Gradient–intercept format, $y = mx + c$
The equation of the line is $y = mx + c$,
where m is the gradient
and c is the y-intercept.

Note:
The y-intercept is the y-coordinate of the point where the line crosses the y-axis.

For example, the line $y = 3x - 2$ has gradient 3 and crosses the y-axis at $(0, -2)$.

The format $ax + by + c = 0$, where a, b and c are integers

Example 2.5 Write the equation of the following straight line in the form $ax + by + c = 0$, where a, b and c are integers.

$$y = -\tfrac{1}{4}x - \tfrac{5}{6}$$

Step 1: Eliminate any fractions.

$(\times 12) \qquad 12y = -3x - 10$

Step 2: Collect all terms on one side, with zero on the other.

$3x + 12y + 10 = 0$

Tip:
To eliminate the fractions multiply *every* term by the common denominator (12).

The format $y - y_1 = m(x - x_1)$
Take P to be the general point (x, y).
A is the fixed point (x_1, y_1).

Using the gradient formula (calling the gradient m):

$$m = \dfrac{y - y_1}{x - x_1}$$

This can be rearranged to give

$$y - y_1 = m(x - x_1)$$

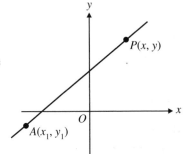

Note:
This form is useful when you know the gradient and a point on the line.

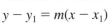

Example 2.6   Find the equation of the line with gradient $\frac{2}{3}$ that passes through the point $(-3, 1)$. Write the answer in the form $ax + by + c = 0$, where a, b and c are integers.

Step 1: Substitute into the formula.

$m = \frac{2}{3}$ and $(x_1, y_1) = (-3, 1)$.

Equation of line is

$$y - y_1 = m(x - x_1)$$

Step 2: Multiply by the denominator if necessary.

$$y - 1 = \frac{2}{3}(x - (-3))$$

$$3(y - 1) = 2(x + 3)$$

Step 3: Rearrange into the required format.

$$3y - 3 = 2x + 6$$

$$-2x + 3y - 9 = 0$$

Note:
You could write
$2x - 3y + 9 = 0$ to avoid
starting with a minus sign.

Example 2.7   The points A and B have coordinates $(12, 5)$ and $(7, 3)$. Find:

a   the gradient of AB

b   the equation of the line AB.

Step 1: Use the gradient formula.

a   Gradient $= \dfrac{3 - 5}{7 - 12} = \dfrac{-2}{-5} = \dfrac{2}{5}$

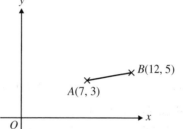

Tip:
It does not matter which point
you use as (x_1, y_1).

Step 2: Use the known gradient, known point formula.

b   Equation of line AB is

$$y - y_1 = m(x - x_1)$$

$$y - 5 = \frac{2}{5}(x - 12)$$

$$5(y - 5) = 2(x - 12)$$

$$5y - 25 = 2x - 24$$

$$5y = 2x + 1$$

You may have to find the **point of intersection** of two lines. In this case, solve the two equations simultaneously.

Example 2.8   Find the point of intersection of the two lines

$y = \frac{2}{3}x - 4$ and $4x - y = 9$.

Step 1: Substitute for one variable from one equation into the other.

$$y = \tfrac{2}{3}x - 4 \qquad ①$$

$$4x - y = 9 \qquad ②$$

Recall:
Simultaneous equations
(Section 1.8).

Substitute for y from ① into ②:

$$4x - (\tfrac{2}{3}x - 4) = 9$$

Step 2: Simplify and solve the resulting equation.

$$4x - \tfrac{2}{3}x + 4 = 9$$

$$4x - \tfrac{2}{3}x = 5$$

Multiply through by 3:

$$12x - 2x = 15$$

$$x = \tfrac{3}{2}$$

Step 3: Substitute the value found into one of the original equations.

Substitute for x in ②:

$$6 - y = 9 \quad \Rightarrow \quad y = -3$$

The lines intersect at $(\frac{3}{2}, -3)$.

1 Write the following lines in the general form $ax + by + c = 0$.

  a $y = 3x - 2$ **b** $x + \frac{2}{3}y = 2$ **c** $\frac{1}{2}x + \frac{3}{4}y = 5$ **d** $\frac{3}{5}x = 2 - \frac{1}{2}y$

2 a Write the straight line $4x - 5y = 8$ in the form $y = mx + c$.

  b What is the gradient and point of intercept on the y-axis of the straight line $2x + 3y = 6$?

3 For the following pairs of points, calculate **i** the gradient of AB **ii** the midpoint of AB **iii** the distance AB, writing your answer in surd form if necessary.

 a $A(-3, 6)$, $B(4, -1)$ **b** $A(4, 6)$, $B(-2, -4)$ **c** $A(-1, -2)$, $B(2, -1)$

4 Find the equation of the line that has:

  a a gradient of $\frac{3}{5}$ passing through the point $(-4, -2)$

  b a gradient of -3 passing through the point $(5, -3)$

  c a line passing through the points $(-3, 6)$ and $(4, -1)$ (Hint: work out the gradient first).

5 A triangle is formed by three straight lines, $y = \frac{1}{2}x$, $2x + y + 5 = 0$ and $x + 3y - 5 = 0$.
Prove that the triangle is isosceles.

6 In the triangle ABC, A, B and C are the points $(-4, 1)$, $(-2, -3)$ and $(3, 2)$, respectively.

  a Show that ABC is isosceles.

  b Find the coordinates of the midpoint of the base.

  c Find the area of ABC.

SKILLS CHECK **2A EXTRA is on the CD**

2.2 Parallel and perpendicular lines

Conditions for two straight lines to be parallel or perpendicular to each other.

For two lines with gradients m_1 and m_2:

The lines are **parallel** if their gradients are the same, i.e. $m_1 = m_2$.

The lines are **perpendicular** if the product of their gradients is -1, i.e. $m_1 \times m_2 = -1$.

Note:
Perpendicular lines have gradients that are negative reciprocals of each other, i.e.
$$m_1 = -\frac{1}{m_2}.$$

Example 2.9 Show that the lines $y = \frac{2}{3}x - 1$ and $2x - 3y + 6 = 0$ are parallel.

Step 1: Rearrange the equations of the lines into the form $y = mx + c$.

The first line has gradient $\frac{2}{3}$.

Rearranging the equation of the second line:

$2x - 3y + 6 = 0$

$\qquad 3y = 2x + 6$

$\qquad\;\; y = \frac{2}{3}x + 2$

The second line has gradient $\frac{2}{3}$.

Recall:
When written in the form $y = mx + c$, the gradient is m.

Step 2: Compare gradients. Since both lines have the same gradient, they are parallel.

Example 2.10   Find the equation of the line passing through (0, 2) that is perpendicular to the line $y = \frac{1}{4}x + 3$.

Step 1: Find the gradient of the required line.

The gradient of the given line is $\frac{1}{4}$.

The gradient of the required line is the negative reciprocal of $\frac{1}{4}$, so the gradient of the required line is -4.

Step 2: Use the gradient and known point to find the equation of the line.

The given point (0, 2) is the intercept on the y-axis, so the equation of the required line is

$y = -4x + 2$.

> **Tip:**
> Unless a particular equation is asked for, you can give any form.

Example 2.11   Find the equation of the perpendicular bisector of the line AB where A is the point $(-3, 5)$ and B is the point $(-1, 3)$.

Step 1: Find the gradient of AB.

Gradient of $AB = \dfrac{y_2 - y_1}{x_2 - x_1} = \dfrac{3 - 5}{-1 - (-3)} = \dfrac{-2}{2} = -1$

Step 2: Find the gradient of a line perpendicular to AB.

Gradient of perpendicular bisector $= 1$

Step 3: Find the midpoint of AB.

Midpoint of AB

$= \left(\dfrac{x_1 + x_2}{2}, \dfrac{y_1 + y_2}{2} \right)$

$= \left(\dfrac{(-3) + (-1)}{2}, \dfrac{5 + 3}{2} \right)$

$= (-2, 4)$

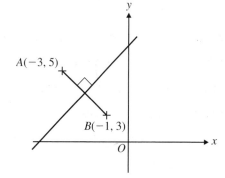

> **Recall:**
> Product of gradients is –1 for perpendicular lines.

> **Tip:**
> It may be helpful to draw a sketch.

Step 4: Find the equation of the line using $y - y_1 = m(x - x_1)$.

The perpendicular bisector has gradient 1 and passes through $(-2, 4)$. The equation of the perpendicular bisector is:

$y - 4 = 1(x - (-2))$

$y - 4 = x + 2$

$y = x + 6$

Example 2.12   The point A has coordinates $(3, -5)$ and the point B has coordinates $(5, 3)$. The midpoint of AB is M and the line MC is perpendicular to AB, where C has coordinates $(8, p)$.

a Find the coordinates of M.

b Find the gradient of MC.

c Find the value of p.

Step 1: Draw a sketch.

Step 2: Use the midpoint formula.

a Midpoint of $AB = \left(\dfrac{3 + 5}{2}, \dfrac{(-5) + 3}{2} \right) = (4, -1)$

Step 3: Use the gradient of the given line to find the gradient of the perpendicular line.

b Gradient of $AB = \left(\dfrac{3 - (-5)}{5 - 3} \right) = \dfrac{8}{2} = 4$

Since AB and MC are perpendicular, the product of the gradients is -1. So the gradient of MC is $-\frac{1}{4}$.

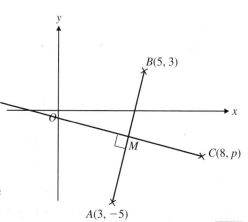

Step 4: Use the gradient formula for the given points.

c Using points $M(4, -1)$ and $C(8, p)$:

gradient of $MC = \dfrac{p - (-1)}{8 - 4} = \dfrac{p + 1}{4}$.

So $\dfrac{p + 1}{4} = -\dfrac{1}{4}$

$p + 1 = -1$

$p = -2$

Example 2.13 The line AB has equation $3x - 2y = 13$ and the points A and B have coordinates $(3, -2)$ and $(5, k)$ respectively.

a Find the value of k.

b Find the equation of the line through A that is perpendicular to AB, giving your answer in the form $ax + by + c = 0$, where a, b and c are integers.

Step 1: Substitute coordinates of B into equation of line.

a Since $(5, k)$ lies on the line $3x - 2y = 13$:

$3 \times 5 - 2k = 13$

$k = 1$

Tip:
If $(5, k)$ lies on the line, then $x = 5$ and $y = k$ must satisfy the equation of the line.

Step 2: Find the gradient of AB.

b A is the point $(3, -2)$, B is the point $(5, 1)$.

Gradient of $AB = \dfrac{1 - (-2)}{5 - 3} = \dfrac{3}{2}$

Tip:
Leave the gradient as a top-heavy fraction.

Step 3: Use the condition for gradients of perpendicular lines.

Gradient of line perpendicular to $AB = -\dfrac{2}{3}$

Equation of line through $A(3, -2)$ with gradient $-\dfrac{2}{3}$ is given by:

Step 4: Use $y - y_1 = m(x - x_1)$.

$y - (-2) = -\dfrac{2}{3}(x - 3)$

$y + 2 = -\dfrac{2}{3}(x - 3)$

Step 5: Rearrange into required format.

$3(y + 2) = -2(x - 3)$

$3y + 6 = -2x + 6$

$2x + 3y = 0$

Tip:
Take care with fractions and negatives when rearranging the equation.

SKILLS CHECK **2B: Parallel and perpendicular lines**

1 State whether the following pairs of lines are parallel, perpendicular or neither:

a $5y = 4x - 7$ **b** $y = \dfrac{2}{3}x - 8$ **c** $3x - 2y + 9 = 0$

$4y = 7 - 5x$ $4x - 6y = 5$ $2x - 3y = 6$

2 Show that the following lines are perpendicular:
$3x - 4y = 20$ and $8x + 6y + 15 = 0$

3 Find the equation of the line parallel to the line $2x - 3y = 6$, passing through the point $(0, 3)$.

4 Find the equation of the line perpendicular to the line $y = \dfrac{4}{5}x - 2$, passing through the point $(0, -2)$.

5 Find the equation of the line perpendicular to $x + 5y - 10 = 0$, passing through the point $(3, -1)$.

6 Find the equation of the perpendicular bisector of AB, where $A(4, -5)$ and $B(-2, -3)$.

7 Find the equation of the line parallel to $y = -1\dfrac{2}{3}x - 2\dfrac{1}{6}$ passing through $(-2, -3)$.

8 Find the equation of the line perpendicular to the line $y = 5x - 8$ and passing through $(2, 2)$.

 9 $A(1, 0)$, $B(3, 5)$ and $C(7, 3)$ are three vertices of a parallelogram $ABCD$. Find:

  a the gradient of AB **b** the equation of CD

  c the gradient of BC **d** the equation of AD

  e the coordinates of D.

SKILLS CHECK **2B EXTRA** is on the CD

Examination practice   Coordinate geometry in the (x, y) plane

1 The straight line l_1 with equation $y = \frac{3}{2}x - 2$ crosses the y-axis at the point P. The point Q has coordinates $(5, -3)$.

  a Calculate the coordinates of the midpoint of PQ.

  The straight line l_2 is perpendicular to l_1 and passes through Q.

  b Find an equation for l_2 in the form $ax + by = c$, where a, b and c are integer constants.

  The lines l_1 and l_2 intersect at the point R.

  c Calculate the exact coordinates of R.   [Edexcel January 2003]

2 a Find an equation of the straight line passing through the points with coordinates $(-1, 5)$ and $(4, -2)$, giving your answer in the form $ax + by + c = 0$, where a, b and c are integers.

  The line crosses the x-axis at the point A and the y-axis at the point B, and O is the origin.

  b Find the area of $\triangle OAB$.   [Edexcel January 1997]

 3 The points A and B have coordinates $(2, 16)$ and $(12, -4)$ respectively. A straight line l_1 passes through A and B.

  a Find an equation for l_1, in the form $ax + by = c$.

  The line l_2 passes through the point C with coordinates $(-1, 1)$ and has gradient $\frac{1}{3}$.

  b Find an equation for l_2.

  The lines l_1 and l_2 intersect at the point D. The point O is the origin.

  c Find the length OD, giving your answer in the form $m\sqrt{5}$, where m is an integer.   [Edexcel January 2000]

4 The straight line l_1 has equation $4y + x = 0$.

  The straight line l_2 has equation $y = 2x - 3$.

  a On the same axes, sketch the graphs of l_1 and l_2. Show clearly the coordinates of all points at which the graphs meet the coordinate axes.

  The lines l_1 and l_2 intersect at the point A.

  b Calculate, as exact fractions, the coordinates of A.

  c Find an equation of the line through A which is perpendicular to l_1. Give your answer in the form $ax + by + c = 0$, where a, b and c are integers.   [Edexcel June 2002]

5 The points $A(-1, -2)$, $B(7, 2)$ and $C(k, 4)$, where k is a constant, are the vertices of $\triangle ABC$. Angle ABC is a right angle.

   a Find the gradient of AB.

   b Calculate the value of k.

   c Show that the length AB may be written in the form $p\sqrt{5}$, where p is an integer to be found.

   d Find the exact value of the area of $\triangle ABC$.

   e Find an equation for the straight line l passing through B and C. Give your answer in the form $ax + by + c = 0$, where a, b and c are integers.

   The line l crosses the x-axis at D and the y-axis at E.

   f Calculate the coordinates of the midpoint of DE.   [Edexcel June 2001]

6 The points A and B have coordinates $(4, 6)$ and $(12, 2)$ respectively.

   The straight line l_1 passes through A and B.

   a Find an equation for l_1 in the form $ax + by = c$, where a, b and c are integers.

   The straight line l_2 passes through the origin and has gradient -4.

   b Write down an equation for l_2.

   The lines l_1 and l_2 intercept at the point C.

   c Find the exact coordinates of the midpoint of AC.   [Edexcel June 2003]

7 The line l has equation $2y + 3x - 6 = 0$ and intersects the coordinate axes at two points $P(p, 0)$ and $Q(0, q)$.

   a Find the values of p and q.

   b Find the coordinates of M, the midpoint of PQ.

   c Find the equation of the line through M perpendicular to PQ, leaving your answer in the form $y = mx + c$.

 8 The point A has coordinates $(3, -5)$ and the point B has coordinates $(1, 1)$.

   a **i** Find the gradient of AB.

   ii Show that the equation of the line AB can be written in the form $rx + y = s$, where r and s are positive integers.

   b The midpoint of AB is M and the line MC is perpendicular to AB.

   i Write down the coordinates of M.

   ii Find the gradient of the line MC.

   iii Given that C has coordinates $(5, p)$, find the value of the constant p.

3 Sequences and series

3.1 Introducing sequences and series

Sequences, including those given by a formula for the nth term.

A sequence is a succession of terms that follow a rule. One way of defining a sequence is to give the formula for the nth term. The sequence is generated by substituting positive integer values of n into the formula.

Note:
The nth term is often denoted by u_n, x_n or t_n.

Example 3.1 The nth term of a sequence is u_n, where $u_n = 2n - 3$ for $n \geqslant 1$. Write down the first four terms of the sequence.

Step 1: Substitute $n = 1, 2, 3$ and 4 into the formula.

$n = 1$ $u_1 = 2 \times 1 - 3 = -1$

$n = 2$ $u_2 = 2 \times 2 - 3 = 1$

$n = 3$ $u_3 = 2 \times 3 - 3 = 3$

$n = 4$ $u_4 = 2 \times 4 - 3 = 5$

The first four terms of the sequence are $-1, 1, 3, 5$.

Note:
This is an example of an arithmetic sequence. See Section 3.3.

Example 3.2 Find the first five terms of the sequence defined by
$$u_n = 3 + \left(\tfrac{1}{2}\right)^n, \quad n \geqslant 1.$$

Step 1: Substitute $n = 1, 2, 3, 4$ and 5 into the formula.

$n = 1$ $u_1 = 3 + \left(\tfrac{1}{2}\right)^1 = 3\tfrac{1}{2}$

$n = 2$ $u_2 = 3 + \left(\tfrac{1}{2}\right)^2 = 3\tfrac{1}{4}$

$n = 3$ $u_3 = 3 + \left(\tfrac{1}{2}\right)^3 = 3\tfrac{1}{8}$

$n = 4$ $u_4 = 3 + \left(\tfrac{1}{2}\right)^4 = 3\tfrac{1}{16}$

$n = 5$ $u_5 = 3 + \left(\tfrac{1}{2}\right)^5 = 3\tfrac{1}{32}$

The sequence is $3\tfrac{1}{2}, 3\tfrac{1}{4}, 3\tfrac{1}{8}, 3\tfrac{1}{16}, 3\tfrac{1}{32}, \ldots$

3.2 Recurrence relations

Sequences generated by a simple relation of the form $x_{n+1} = f(x_n)$.

Another way of defining a sequence is to give a **recurrence relation**. This could be in the form $x_{n+1} = f(x_n)$, where a subsequent term is given in relation to the previous term. To generate the sequence, you must know the first term, x_1.

Note:
This is also known as an **iterative formula**.

Example 3.3 A sequence is defined by the recurrence relation $x_{n+1} = 3x_n - 2$, with $x_1 = 4$. Calculate x_2, x_3 and x_4.

Step 1: Substitute $n = 1, 2$ and 3 into the formula.

$n = 1$ $x_2 = 3x_1 - 2 = 3 \times 4 - 2 = 10$

$n = 2$ $x_3 = 3x_2 - 2 = 3 \times 10 - 2 = 28$

$n = 3$ $x_4 = 3x_3 - 2 = 3 \times 28 - 2 = 82$

Example 3.4 A sequence has recurrence relation $x_{n+1} = \dfrac{4}{x_n} + 4$, $x_1 = 1$.

Calculate x_2, x_3 and x_4.

Step 1: Substitute
$n = 1, 2, 3, \ldots$
successively into the
formula.

$x_1 = 1$

$x_2 = \dfrac{4}{1} + 4 = 8$

$x_3 = \dfrac{4}{8} + 4 = 4\tfrac{1}{2} = \tfrac{9}{2}$

$x_4 = \dfrac{4}{\frac{9}{2}} + 4 = 4 \times \tfrac{2}{9} + 4 = 4\tfrac{8}{9} = \tfrac{44}{9}$

Series

When the terms of a sequence are added, a **series** is formed, for example $2 + 6 + 10 + 14 + 18 + 22$.

There is a shorthand way of writing the terms of a series, using a general term and Σ (sigma) notation.

> **Note:**
> Σ means 'the sum of' and is read 'sigma'.

For example, to evaluate $\displaystyle\sum_{r=1}^{5} (2r + 3)$, add the terms calculated by substituting $r = 1, 2, 3, 4$ and 5.

When $r = 1$, $2r + 3 = 5$; when $r = 2$, $2r + 3 = 7$, and so on.

So $\displaystyle\sum_{r=1}^{5} (2r + 3) = 5 + 7 + 9 + 11 + 13 = 45$.

Example 3.5 Find **a** $\displaystyle\sum_{r=1}^{4} (3r - 2)$ **b** $\displaystyle\sum_{r=4}^{7} \dfrac{2r - 1}{r + 1}$.

> **Note:**
> Letters other than r may be used.

Step 1: Substitute
$r = 1, 2, 3, 4$.

a $\displaystyle\sum_{r=1}^{4} (3r - 2) = (3(1) - 2) + (3(2) - 2) + (3(3) - 2) + (3(4) - 2)$

$\qquad\qquad\qquad\qquad \uparrow \qquad\qquad \uparrow \qquad\qquad \uparrow \qquad\qquad \uparrow$

$\qquad\qquad\qquad\quad (r = 1) \quad\;\; (r = 2) \quad\;\; (r = 3) \quad\;\; (r = 4)$

Step 2: Add the terms.

$\displaystyle\sum_{r=1}^{4} (3r - 2) = 1 + 4 + 7 + 10$

$\qquad\qquad\quad = 22$

> **Note:**
> This is an arithmetic series (Section 3.3).

Step 1: Substitute
$r = 4, 5, 6, 7$.

b $\displaystyle\sum_{r=4}^{7} \dfrac{2r - 1}{r + 1} = \dfrac{2(4) - 1}{4 + 1} + \dfrac{2(5) - 1}{5 + 1} + \dfrac{2(6) - 1}{6 + 1} + \dfrac{2(7) - 1}{7 + 1}$

Step 2: Add the terms.

$\qquad = \dfrac{7}{5} + \dfrac{9}{6} + \dfrac{11}{7} + \dfrac{13}{8} = \dfrac{1707}{280}$

1 The nth term, u_n of a sequence is $n^2 - 1$. Find the first four terms.

2 If $x_n = 1 - \left(\frac{2}{3}\right)^n$, find x_1, x_2, x_3 and x_4.

3 Calculate the next four terms of the sequence defined by the recurrence relation

   a $x_{n+1} = 2x_n, \ x_1 = 2$ **b** $x_{n+1} = 2x_n + 1, \ x_1 = 3.$

 4 A sequence has a recurrence relation $x_{n+1} = \dfrac{5 - 4x_n}{x_n}, \ x_1 = -1.$

   Calculate x_2 and x_3.

5 Evaluate $\displaystyle\sum_{r=1}^{5} u_r$ where

   a $u_r = 3 + 4r$ **b** $u_r = 0.5(2^r)$

6 Evaluate

   a $\displaystyle\sum_{r=1}^{3} (20 - 3r)$ **b** $\displaystyle\sum_{r=1}^{4} 2^{r-1}$

7 A recurrence relation is defined by $u_{r+2} = 2u_{r+1} - 3u_r - 1$, where $u_1 = 1, u_2 = 0$. Find u_3, u_4 and u_5.

8 Evaluate

   a $\displaystyle\sum_{r=1}^{5} (r^2 - r)$ **b** $\displaystyle\sum_{r=4}^{6} \dfrac{r}{r+1}$

SKILLS CHECK **3A EXTRA** is on the CD

3.3 Arithmetic series

Arithmetic series, including the formula for the first n natural numbers.

The sequence defined by $u_n = 4n + 1$ is 5, 9, 13, 17, 21, ...

This is an example of an **arithmetic sequence**, where each successive term is obtained from the previous one by *adding* a constant amount, called the **common difference**.

Adding the terms of an arithmetic sequence gives an **arithmetic series**, for example $5 + 9 + 13 + 17 + 21 + \ldots$

Note:
The iterative formula for this sequence is $x_{n+1} = x_n + 4$, with $x_1 = 5$.

General expression for u_n

In an arithmetic series, the first term is usually denoted by a and the common difference is denoted by d.

The terms are as follows:

$$\text{1st term} = a$$
$$\text{2nd term} = a + d$$
$$\text{3rd term} = (a + d) + d = a + 2d$$
$$\text{4th term} = (a + 2d) + d = a + 3d$$

Continuing the pattern,

$$n\text{th term} = a + (n - 1)d$$

In the series $5 + 9 + 13 + 17 + 21 + \ldots$, $a = 5$ and $d = 4$,

so the nth term $= 5 + (n - 1) \times 4$
$$= 5 + 4n - 4$$
$$= 4n + 1, \text{ as expected.}$$

Recall:
Each time we add d to get the next term.

Tip:
The formula for the nth term will be given in the examination but it is useful to learn it.

Note:
Notice that the formula for the nth term is linear and the coefficient of n gives the value of d.

Example 3.6 The third term in an arithmetic series is 8 and the eighth term is 24. Find

a the first term and the common difference,

b the 23rd term.

Step 1: Identify the type of series and write the general series.

Step 2: Underneath write the information given in the question.

Step 3: Find any unknowns by solving the equations.

a This is an arithmetic series.
The general series is

$$a + \quad a + d \quad + \quad a + 2d \quad + \cdots + \quad a + 7d \quad + \cdots + \quad a + (n - 1)d$$
$$\cdots + 8 \qquad\qquad + \cdots + 24 \qquad\qquad + \cdots$$

$$a + 2d = 8 \quad ①$$
$$a + 7d = 24 \quad ②$$
$$② - ①: 5d = 16 \Rightarrow d = 3.2$$

Substitute into ①: $a = 8 - 2d = 8 - 2 \times 3.2 = 1.6$

The first term is 1.6 and the common difference is 3.2.

Step 4: Use an appropriate formula.

b nth term $= a + (n - 1)d$

So the 23rd term $= 1.6 + 22 \times 3.2 = 72$

Tip:
Ensure that you include the relevant terms given in the question.

Recall:
Solving simultaneous equations (Section 1.8).

Sum of first n terms, S_n

The sum of the first n terms of an arithmetic series, with first term a and common difference d, is written S_n and is given by

$$S_n = \frac{n}{2}\{2a + (n - 1)d\}$$

Alternatively, if you know the first term and the last term, l, use

$$S_n = \frac{n}{2}(a + l)$$

Tip:
These formulae will be given in the examination but you must learn how to prove them.

Tip:
$l = a + (n - 1)d$

Proof of $S_n = \frac{n}{2}\{2a + (n-1)d\}$ and $S_n = \frac{n}{2}(a + l)$

Step 1: Write the general series.

The sum of the first n terms is given by:

$$S_n = \quad a \quad + \quad (a+d) \quad + \cdots + \{a + (n-2)d\} + \{a + (n-1)d\}$$

Step 2: Rewrite the series in reverse order, with the terms corresponding with the original sum.

$$S_n = \{a + (n-1)d\} + \{a + (n-2)d\} + \cdots + (a+d) \quad + \quad a$$

Adding

Step 3: Add the two sums by adding corresponding terms.

$$2S_n = \{2a + (n-1)d\} + \{2a + (n-1)d\} + \cdots + \{2a + (n-1)d\} + \{2a + (n-1)d\}$$
$$= n\{2a + (n-1)d\}$$

Step 4: Divide by 2.

Dividing by 2:

$$S_n = \frac{n}{2}\{2a + (n-1)d\}, \text{ as required.}$$

> **Note:**
> The term $\{2a + (n-1)d\}$ occurs n times, as there are n terms.

Remember the nth term is given by $a + (n-1)d$ and so the last term, l, of a series with n terms is also given by $l = a + (n-1)d$. Therefore

$$S_n = \frac{n}{2}\{2a + (n-1)d\} = \frac{n}{2}\{a + a + (n-1)d\}$$

$$S_n = \frac{n}{2}(a + l)$$

You can use this formula for the sum of an arithmetic series if you know the first term and the last term, l.

Example 3.7 **a** Using the above formula, find the sum of the first five terms of the arithmetic series

$$\tfrac{3}{2} + \tfrac{5}{6} + \tfrac{1}{6} - \tfrac{1}{2} - \cdots$$

b Verify this sum by adding the individual terms of the series.

Step 1: Identify the type of series and write the general series.

a You are told this is an arithmetic series.
The general series is:

$$a \quad + \quad a+d \quad + \quad a+2d \quad + \quad a+3d + \cdots + \quad a + (n-1)d$$
$$\tfrac{3}{2} \quad + \quad \tfrac{5}{6} \quad + \quad \tfrac{1}{6} \quad + \quad -\tfrac{1}{2} + \cdots$$

Step 2: Underneath write the information given in the question.

$$a = \tfrac{3}{2}; a + d = \tfrac{5}{6} \Rightarrow d = \tfrac{5}{6} - \tfrac{3}{2} = -\tfrac{2}{3}$$

Step 3: Find any unknowns by solving the equations.

$$S_n = \frac{n}{2}\{2a + (n-1)d\}, \text{ with } n = 5$$

Step 4: Use an appropriate formula.

$$S_5 = \tfrac{5}{2}\{2 \times \tfrac{3}{2} + (5-1)(-\tfrac{2}{3})\} = \tfrac{5}{6}$$

b Calculating the sum, we get $\tfrac{3}{2} + \tfrac{5}{6} + \tfrac{1}{6} - \tfrac{1}{2} - \tfrac{7}{6} = \tfrac{5}{6}$, verifying the answer in part **a**.

> **Note:**
> 5th term $= -\tfrac{1}{2} - \tfrac{2}{3} = -\tfrac{7}{6}$.

Example 3.8 The eighth term of an arithmetic series is twice the fourth term, and the first term is 6. Find the sum of the first 25 terms of the series.

Step 1: Identify the type of series and write the general series.

This is an arithmetic series.

$$a \quad + \quad (a+d) \quad + \quad (a+2d) \quad + \quad (a+3d) \quad + \cdots + \quad (a+7d) \quad + \cdots + \quad (a + (n-1))$$
$$\downarrow \qquad\qquad\qquad\qquad\qquad\qquad\qquad\qquad\qquad\qquad\qquad\qquad \downarrow$$
$$6 \qquad\qquad\qquad\qquad\qquad\qquad\qquad\qquad\qquad\qquad\qquad\qquad + 2(a+3d)$$
$$\qquad\qquad\qquad\qquad\qquad\qquad\qquad\qquad\qquad\qquad\qquad\qquad\qquad \downarrow$$
$$\qquad\qquad\qquad\qquad\qquad\qquad\qquad\qquad\qquad\qquad\qquad\qquad\qquad \text{4th term}$$

Step 2: Underneath write the information given in the question.

$$a = 6$$

> **Tip:**
> The 8th term $= 2(4\text{th term})$.

Step 3: Find any unknowns by solving the equations.

$$a + 7d = 2(a + 3d)$$
$$\Rightarrow 6 + 7d = 2(6 + 3d)$$
$$6 + 7d = 12 + 6d$$
$$d = 6$$

Step 4: Use an appropriate formula.

$$S_n = \frac{n}{2}\{2a + (n-1)d\}, n = 25$$

$$S_{25} = \tfrac{25}{2}\{2 \times 6 + (25-1)6\} = 1950$$

Example 3.9 Ben's aunt gave him money on his birthday every year from his 15th birthday to his 30th birthday.

She gave him £200 on his 15th birthday. How much did she give him in total if on each subsequent year she gave him

  a £100,

  b £100 more than on his previous birthday?

Step 1: Identify the type of sequence and write the general sequence.
Step 2: Underneath write the information given in the question.
Step 3: Find any unknowns by solving the equations.
Step 4: Use an appropriate formula.

a This is an arithmetic sequence. The general sequence is

$$a, \quad a + d, \quad a + 2d, \quad a + 3d, \quad \ldots, \quad a + (n - 1)d$$
$$200 \quad\quad 300 \quad\quad\quad 400 \quad\quad\quad 500$$

The total amount Ben had received by his 30th birthday is given by the 16th term of the sequence.

nth term $= a + (n - 1)d$

16th term $= 200 + 15 \times 100 = 1700$

Ben had been given £1700.

b In this part the total amount given to Ben is the sum of the first 16 terms of the series $200 + 300 + 400 + \cdots$

$$S_n = \frac{n}{2}\{2a + (n - 1)d\}$$

$$S_{16} = \frac{16}{2}(2 \times 200 + 15 \times 100) = 15\,200$$

Ben had been given £15 200.

Using Σ notation to describe an arithmetic series

The nth term in an arithmetic sequence can be written in the form $dn + c$, where d is the common difference and c is a constant. Hence an arithmetic series, with common difference d, can be written

$$\sum_{r=1}^{n} (dr + c).$$

For example, $\displaystyle\sum_{r=1}^{5} (4r + 1) = 5 + 9 + 13 + 17 + 21 = 65$.

Example 3.10 Evaluate $\displaystyle\sum_{r=1}^{14} (3r - 2)$.

Step 1: Substitute integer values of r, from the lower to higher number.
Step 2: Identify the series and use an appropriate formula.

$$\sum_{r=1}^{14} (3r - 2) = 1 + 4 + 7 + \cdots + 40$$

This is an arithmetic series with $a = 1, d = 3, l = 40, n = 14$.

$$\sum_{r=1}^{14} (3r - 2) = S_{14} = \frac{14}{2}(1 + 40) = 287$$

Tip:
In **a**, each term is the total amount Ben had been given on and before the particular birthday.

Tip:
Take care when working out n.

Tip:
In **b**, the total amount is found by adding the amounts given each year.

Recall:
Σ means 'the sum of' (see Section 3.2).

Note:
Σ notation can be used for other series, such as geometric series studied in C2.

Note:
The series is arithmetic, with common difference 4.

Tip:
Work out the first few terms and the last term.

Tip:
Use $S_n = \dfrac{n}{2}(a + l)$.

Example 3.11 Show that $\displaystyle\sum_{r=1}^{n} r = \frac{n(n+1)}{2}$.

Step 1: Substitute integer values of r, from the lower to higher number.

$$\sum_{r=1}^{n} r = 1 + 2 + 3 + \cdots + n$$

Step 2: Identify the type of series.

This is an arithmetic series, with n terms, where $a = 1$, $d = 1$, $l = n$.

Step 3: Use an appropriate formula.

$$S_n = \frac{n}{2}(a + l) = \frac{n}{2}(1 + n) = \frac{n(n+1)}{2}$$

Note:
You could use
$S_n = \frac{n}{2}\{2a + (n-1)d\}$

The result in Example 3.12 is very important and it is helpful to learn it.

The **sum of the first n natural numbers** is given by the following:

$$\sum_{r=1}^{n} r = \frac{n(n+1)}{2}$$

Note:
Natural numbers are the counting numbers 1, 2, 3, 4, …

Example 3.12 Evaluate **a** $\displaystyle\sum_{r=1}^{100} r$ **b** $\displaystyle\sum_{r=20}^{50} r$.

Step 1: Use the Σr formula.

a $\displaystyle\sum_{r=1}^{100} r = 1 + 2 + 3 + \cdots + 100 = \frac{100 \times 101}{2} = 5050$

Step 1: Use the Σr formula in two stages, subtracting unwanted terms.

b $\displaystyle\sum_{r=20}^{50} r = \sum_{r=1}^{50} r - \sum_{r=1}^{19} r$

$$= \frac{50 \times 51}{2} - \frac{19 \times 20}{2}$$
$$= 1275 - 190$$
$$= 1085$$

Tip:
The series is
$20 + 21 + \cdots + 50$.

Tip:
To use the Σr formula, the lowest value of r must be 1.

The following relationship can be used to sum any arithmetic series:

$$\sum_{r=1}^{n} (dr + c) = d\sum_{r=1}^{n} r + \sum_{r=1}^{n} c = d\sum_{r=1}^{n} r + nc$$

Note:
$\displaystyle\sum_{r=1}^{n} c = c + \cdots + c = nc$

Example 3.13 Evaluate $\displaystyle\sum_{r=1}^{24} (3r + 2)$.

Step 1: Expand, using Σ notation.

$$\sum_{r=1}^{24} (3r + 2) = 3\sum_{1}^{24} r + 24 \times 2$$

Step 2: Apply the formula for Σr.

$$= 3 \times \frac{24 \times 25}{2} + 48$$
$$= 948$$

Tip:
Choose the method you prefer.

Alternatively, you can use the expanded series and the formula for the sum of an arithmetic series:

Step 1: Write out the first few and the last terms.

$$\sum_{r=1}^{24} (3r + 2) = 5 + 8 + 11 + \cdots + 74$$

Tip:
Substitute $r = 1, 2, 3, \ldots, 24$.

Step 2: Identify the series.

This is an arithmetic series with $a = 5$, $d = 3$, $l = 74$, $n = 24$.

Step 3: Use the appropriate formula.

$$S_n = \frac{n}{2}(a + l)$$
$$= \frac{24}{2}(5 + 74)$$
$$= 948$$

1 Find the common difference and the sum of the first 20 terms of the following series:

   a $12 + 17 + 22 + \cdots$ **b** $-2 - 5 - 8 - \cdots$

2 Find the nth term and the sum of the first 200 terms of the arithmetic series $\frac{1}{2} + \frac{3}{2} + \frac{5}{2} + \frac{7}{2} + \cdots$

3 The first term of an arithmetic series is 8 and the seventh term is 26. Find

   a the common difference,   **b** the nth term,   **c** the sum of the first 25 terms.

 4 The third, fourth and fifth terms of an arithmetic series are $(4 + x)$, $2x$ and $(8 - x)$ respectively.

   a Find the value of x.

   b Find the first term, a, and the common difference, d.

   c Find the sum of the first 30 terms of the series.

5 The cost to a company of training a student is £1000 for the first student. The cost is then reduced by £50 for the second student, by a further £50 for the third student and so on, so that the cost for the second student is £950, for the third student is £900 and so on.

   a How much will it cost to train the 20th student?

   b How much will it cost the company to train 20 students?

6 Evaluate   **a** $\displaystyle\sum_{r=1}^{38} r$ **b** $\displaystyle\sum_{r=1}^{18} (7r - 1)$.

7 Evaluate $\displaystyle\sum_{r=12}^{20} (\tfrac{1}{2}r + 3)$.

 8 Find n if $\displaystyle\sum_{r=1}^{2n} (4r - 1) = \sum_{r=1}^{n} (3r + 59)$.

9 The nth term of an arithmetic sequence is u_n, where $u_n = 6 + 2n$.

   a Find the values of u_1, u_2 and u_3.

   b Write down the common difference of the arithmetic sequence.

   c Find the value of n for which $u_n = 46$.

   d Evaluate $\displaystyle\sum_{n=1}^{20} u_n$.

10 Sharon borrowed £5625 on an interest-free loan. She paid back £25 at the end of the first month, then increased her payment by £50 in each subsequent month, paying £75 at the end of the second month, £125 at the end of the third month and so on.

   a How many months did she take to pay off the loan?

   b What was the amount of her final month's repayment?

11 The sum of the first 15 terms of an arithmetic series is 90, and the sum of the first 30 terms is -270. Find the first term, the common difference and the sum of the first 45 terms.

SKILLS CHECK **3B EXTRA is on the CD**

1 The sum of the first two terms of an arithmetic series is 47. The thirtieth term of this series is −62. Find

  a the first term of the series and the common difference,

  b the sum of the first 60 terms of the series.   [Edexcel June 1996]

2 The sum of an arithmetic series is:

$$\sum_{r=1}^{n} (80 - 3r).$$

  a Write down the first two terms of the series.

  b Find the common difference of the series.

  Given that $n = 50$,

  c find the sum of the series.   [Edexcel November 2003]

3 A polygon has 10 sides. The lengths of the sides, starting with the smallest, form an arithmetic series. The perimeter of the polygon is 675 cm and the length of the longest side is twice that of the shortest side. Find, for this series,

  a the common difference,

  b the first term.   [Edexcel January 1998]

4 A sequence of numbers $\{u_n\}$ is defined for $n \geq 1$, by the recurrence relation $u_{n+1} = ku_n - 3$, where k is a constant. Given that $u_1 = 2$,

  a find expressions, in terms of k, for u_2 and u_3.

  Given also that $u_3 = 17$,

  b use algebra to find the possible values of k.

  Given also that $k > 0$,

  c calculate the value of u_4.

5 The fifth term of an arithmetic series is 14 and the sum of the first three terms of the series is −3.

  a Use algebra to show that the first term of the series is −6 and calculate the common difference of the series.

  Given that the nth term of the series is greater than 282,

  b find the least possible value of n.   [Edexcel June 2001]

6 The fourth term of an arithmetic series is $3k$, where k is a constant, and the sum of the first six terms of the series is $7k + 9$.

  a Show that the first term of the series is $9 - 8k$.

  b Find an expression for the common difference of the series, in terms of k.

  Given that the seventh term of the series is 12, calculate

  c the value of k,

  d the sum of the first 20 terms of the series.   [Edexcel January 2001]

7 A string is divided into ten pieces whose lengths are in arithmetic progression. The length of the longest piece is seven times the length of the shortest piece. Given also that the total length of the string is 120 cm, find the length of the shortest piece.

8 The sum of the first seven terms of an arithmetic series is 0 and the fifth term of the series is 2.

  a Find the first term and the common difference of the series.

  b Find the sum of the first 20 terms of the series.

 9 Find n if

$$\sum_{r=1}^{n} (3r + 9) = \sum_{r=1}^{2n} (2r - 2)$$

10 A sequence of numbers $\{u_n\}$ is defined, for $n \geqslant 1$, by the recurrence relation

$$u_{n+1} = 4u_n + k,$$

where k is a constant. Given that $u_1 = 1$,

  a find u_2 and u_3 in terms of k.

Given also that u_3 is three times the value of u_2,

  b find the value of k

  c find u_4 and u_5.

4 Differentiation

4.1 Differentiation, gradients and rates of change

The derivative of f(x) as the gradient of the tangent to the graph of y = f(x) at a point; the gradient of the tangent as a limit; interpretation as a rate of change.

On the curve $y = f(x)$, the slope may vary for different values of x.

At a point P on the curve, a measure of the slope is the **gradient** at P. This tells you how y is changing in relation to x at that point. This is described as the **rate of change** of y with respect to x.

The **gradient of the curve** at P is defined as the **gradient of the tangent** at P. The tangent is the limiting position of the line through P and point Q on the curve as Q gets closer and closer to P.

Note:
The tangent at P *touches* the curve at P.

The gradient of the tangent at a point can be found by **differentiating** $y = f(x)$.

Tip:
See Tangents and normals (Section 4.4).

This gives the **gradient function**, also known as the **derived function** or the **derivative** with respect to x.

The gradient function is usually written $\dfrac{dy}{dx}$ or $f'(x)$.

Note:
$\dfrac{dy}{dx}$ is read as 'dee y by dee x' and $f'(x)$ is read as 'f dashed x'.

4.2 Differentiation of x^n

Differentiation of x^n and related sums and differences.

Differentiate powers of x according to the following rule:

$$y = x^n \Rightarrow \frac{dy}{dx} = nx^{n-1}$$

where n is a rational number.

Tip:
Multiply by the power of x and decrease the power by 1.

If a is a constant:

$$y = ax^n \Rightarrow \frac{dy}{dx} = nax^{n-1}$$

It is useful to remember the following:

$$y = ax \Rightarrow \frac{dy}{dx} = a$$

$$y = a \Rightarrow \frac{dy}{dx} = 0$$

Note:
In function notation, $f(x) = ax^n$
$\Rightarrow f'(x) = nax^{n-1}$.

Note:
$y = ax$ is a line, with constant gradient.
$y = a$ is a line parallel to the x-axis, with zero gradient.

Example 4.1 Find $\dfrac{dy}{dx}$ when   **a** $y = x^3$ **b** $y = 4x^2$ **c** $y = 5$.

Step 1: Use the differentiation rule.

a $\quad y = x^3$

$\dfrac{dy}{dx} = 3x^2$

b $\quad y = 4x^2$

$\dfrac{dy}{dx} = 2 \times 4x^1 = 8x$

c $\quad y = 5$

$\dfrac{dy}{dx} = 0$

To differentiate an expression in x containing several terms, for example $3x^7 - 2x^3 + 3x - 1$, differentiate the terms individually, using the rule:

$$y = f(x) \pm g(x) \Rightarrow \frac{dy}{dx} = f'(x) \pm g'(x)$$

Example 4.2 It is given that $y = 3x^7 - 2x^3 + 3x - 1$. Find the derivative of y with respect to x.

Step 1: Differentiate term by term to get $\frac{dy}{dx}$.

$y = 3x^7 - 2x^3 + 3x - 1$

$\frac{dy}{dx} = 21x^6 - 6x^2 + 3$

Example 4.3 Find $f'(x)$ where $f(x) = (2x - 3)^2$.

Step 1: Write terms in the form ax^n using the index laws.

$f(x) = (2x - 3)^2$
$\quad\ = 4x^2 - 12x + 9$

Step 2: Differentiate term by term to get $f'(x)$.

$f'(x) = 8x - 12$

Example 4.4 Differentiate with respect to x:

a $y = \dfrac{1}{x^2}$ **b** $y = 3\sqrt{x} + \dfrac{2}{x^4}$

Step 1: Write terms in the form ax^n using the index laws.

Step 2: Differentiate term by term to get $\frac{dy}{dx}$.

a $y = \dfrac{1}{x^2} = x^{-2}$

$\dfrac{dy}{dx} = -2x^{-3}$

b $y = 3\sqrt{x} + \dfrac{2}{x^4} = 3x^{\frac{1}{2}} + 2x^{-4}$

$\dfrac{dy}{dx} = 3 \times \tfrac{1}{2}x^{-\frac{1}{2}} + 2 \times (-4)x^{-5}$

$\quad\ = \tfrac{3}{2}x^{-\frac{1}{2}} - 8x^{-5}$

$\quad\ = \dfrac{3}{2\sqrt{x}} - \dfrac{8}{x^5}$

Example 4.5 Differentiate $y = \dfrac{6x^2 + x^3 - 2x}{2x}$ with respect to x.

Step 1: Write terms in the form ax^n using the index laws.

$y = \dfrac{6x^2 + x^3 - 2x}{2x}$

$\quad\ = \dfrac{6x^2}{2x} + \dfrac{x^3}{2x} - \dfrac{2x}{2x}$

$\quad\ = 3x + \tfrac{1}{2}x^2 - 1$

Step 2: Differentiate term by term to get $\frac{dy}{dx}$.

$\dfrac{dy}{dx} = 3 + x$

Example 4.6 Given that $y = \dfrac{x^2 + x}{\sqrt{x}}$, show that $\dfrac{dy}{dx} = \dfrac{3x + 1}{2\sqrt{x}}$.

Step 1: Write terms in the form ax^n using the index laws.

$$y = \frac{x^2 + x}{\sqrt{x}}$$

$$= \frac{x^2}{\sqrt{x}} + \frac{x}{\sqrt{x}}$$

$$= \frac{x^2}{x^{\frac{1}{2}}} + \frac{x}{x^{\frac{1}{2}}}$$

$$= x^{\frac{3}{2}} + x^{\frac{1}{2}}$$

Step 2: Differentiate term by term to get $\dfrac{dy}{dx}$.

$$\frac{dy}{dx} = \tfrac{3}{2}x^{\frac{1}{2}} + \tfrac{1}{2}x^{-\frac{1}{2}}$$

Step 3: Simplify to the required form.

$$= \frac{3\sqrt{x}}{2} + \frac{1}{2\sqrt{x}}$$

$$= \frac{3\sqrt{x}}{2} \times \frac{\sqrt{x}}{\sqrt{x}} + \frac{1}{2\sqrt{x}}$$

$$= \frac{3x}{2\sqrt{x}} + \frac{1}{2\sqrt{x}}$$

$$= \frac{3x + 1}{2\sqrt{x}}$$

> **Tip:**
> Divide each term in the numerator by $\sqrt{x}$.

> **Recall:**
> $\dfrac{1}{\sqrt{x}} = \dfrac{1}{x^{\frac{1}{2}}}$

> **Tip:**
> It is often easier to simplify terms involving square roots and negative indices when they are not written in index form.

4.3 Differentiation, gradients and rates of change

Applications of differentiation to gradients.

To find the gradient at a point on a curve, substitute the x-value of the point into the gradient function.

Example 4.7 The sketch shows the curve $y = (x - 3)^2$.
Using differentiation, find the gradient when

a $x = 3$,

b $x = 4$.

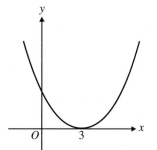

Step 1: Write terms in the form ax^n using the index laws.

$$y = (x - 3)^2$$

$$= x^2 - 6x + 9$$

Step 2: Differentiate term by term to get $\dfrac{dy}{dx}$.

$$\frac{dy}{dx} = 2x - 6$$

Step 3: Substitute the x-value into $\dfrac{dy}{dx}$.

a At $x = 3$, $\dfrac{dy}{dx} = 2(3) - 6 = 0$.

The gradient when $x = 3$ is 0.

b At $x = 4$, $\dfrac{dy}{dx} = 2(4) - 6 = 2$.

The gradient when $x = 4$ is 2.

> **Tip:**
> Notice that the tangent at $x = 3$ is horizontal, so you expect the gradient to be zero. Differentiating provides a check.

> **Tip:**
> Expand the bracket.

Example 4.8 A curve has equation $y = 2\sqrt{x} - 2x$.

Differentiate y with respect to x and hence find the gradient at the point $(4, -4)$.

Step 1: Write terms in the form ax^n using the index laws.

Step 2: Differentiate term by term to get $\dfrac{dy}{dx}$.

$$y = 2\sqrt{x} - 2x = 2x^{\frac{1}{2}} - 2x$$

$$\frac{dy}{dx} = 2 \times \tfrac{1}{2}x^{-\frac{1}{2}} - 2$$

$$= x^{-\frac{1}{2}} - 2$$

$$= \frac{1}{\sqrt{x}} - 2$$

Recall:
To find the gradient at a point, substitute the x-value into the gradient function, $\dfrac{dy}{dx}$ (Section 4.1).

Step 3: Substitute the x-value into $\dfrac{dy}{dx}$.

When $x = 4$, $\dfrac{dy}{dx} = \dfrac{1}{\sqrt{4}} - 2 = -\tfrac{3}{2}$.

The gradient at $(4, -4)$ is $-\tfrac{3}{2}$.

Example 4.9 It is given that $f(x) = \left(\dfrac{1}{x} - \dfrac{1}{x^2}\right)^2$.

a Find the derived function, $f'(x)$.

b Find $f'(1)$ and hence write down the gradient of the curve $y = f(x)$ at the point $(1, 0)$.

a $f(x) = \left(\dfrac{1}{x} - \dfrac{1}{x^2}\right)^2$

Step 1: Write terms in the form ax^n using the index laws.

$$= (x^{-1} - x^{-2})^2$$

$$= x^{-2} - 2x^{-3} + x^{-4}$$

Step 2: Differentiate term by term to get $\dfrac{dy}{dx}$.

$$f'(x) = -2x^{-3} - 2(-3)x^{-4} + (-4)x^{-5}$$

$$= -2x^{-3} + 6x^{-4} - 4x^{-5}$$

Recall:
$(a - b)^2 = a^2 - 2ab + b^2$

Recall:
Index laws (Section 1.1).

Step 3: Substitute the x-value into $f'(x)$.

b $f'(1) = -2(1)^{-3} + 6(1)^{-4} - 4(1)^{-5}$

$$= -2 + 6 - 4$$

$$= 0$$

Note:
You could write
$$f'(x) = -\frac{2}{x^3} + \frac{6}{x^4} - \frac{4}{x^5}$$

The gradient at $(1, 0)$ is given by $f'(1)$.

Hence the gradient at $(1, 0)$ is zero.

Second-order derivatives

The second-order derivative, $\dfrac{d^2y}{dx^2}$, is obtained by differentiating $\dfrac{dy}{dx}$ with respect to x.

In function notation, if $y = f(x)$, the second-order derivative is written $f''(x)$.

Example 4.10 Given that $y = 4x^3 + 5x^2 - 3x + 1$, find **a** $\dfrac{dy}{dx}$ **b** $\dfrac{d^2y}{dx^2}$.

$$y = 4x^3 + 5x^2 - 3x + 1$$

Step 1: Find $\dfrac{dy}{dx}$.

a $\dfrac{dy}{dx} = 12x^2 + 10x - 3$

Step 2: Find $\dfrac{d^2y}{dx^2}$ by differentiating again.

b $\dfrac{d^2y}{dx^2} = 24x + 10$

Example 4.11 If $f(x) = x^3$, find $f''(2)$.

$$f(x) = x^3$$

Step 1: Find $f'(x)$. $\qquad f'(x) = 3x^2$
Step 2: Find $f''(x)$ by $\qquad f''(x) = 6x$
differentiating again.
Step 3: Substitute $x = 2$. $\quad \Rightarrow \quad f''(2) = 6 \times 2 = 12$

Differentiation as a rate of change

$\dfrac{dy}{dx}$ can also be described as the rate of change of y with respect to x. If

you substitute $x = a$ into $\dfrac{dy}{dx}$, then this tells you the rate of change of y

with respect to x at the point where $x = a$.

Recall:
The gradient at a given point is the rate of change of y with respect to x.

Example 4.12 A circular oil slick gets larger with time. The radius of the oil slick x cm at time t seconds is given by $x = 2t^3$.

  a Find the rate of change of x with respect to time at the instant when $t = 3$.

  b Express A cm^2, the area of the oil slick at time t seconds, in terms of t and so find the rate of change of A with respect to t when $t = 1$.

Step 1: Differentiate to find $\dfrac{dx}{dt}$.

a $\quad x = 2t^3$

$$\frac{dx}{dt} = 6t^2$$

Step 2: Substitute the t-value into $\dfrac{dx}{dt}$.

At $t = 3$, $\dfrac{dx}{dt} = 6(3)^2 = 54$.

The rate of change of the radius with time at 3 seconds is 54 cm/s.

Note:
You are finding the change in radius (cm) per second, and so the units for this rate of change are cm/s (or cm s^{-1}).

Step 1: Find a formula relating the measure that is required to the given measure.

b Area of the circle, $A = \pi x^2$

Substitute $x = 2t^3$: $\quad A = \pi(2t^3)^2$
$$= 4\pi t^6$$

Tip:
You need to express A in terms of t.

Step 2: Differentiate with respect to t to get $\dfrac{dA}{dt}$.

$$\frac{dA}{dt} = 24\pi t^5$$

Step 3: Substitute the t-value into $\dfrac{dA}{dt}$.

At $t = 1$, $\dfrac{dA}{dt} = 24\pi(1)^5 = 24\pi$.

The rate of change of the area with time at 1 second is 24π cm^2/s.

Note:
If the rate of change is negative then this means that the variable decreases with respect to time.

Example 4.13 A rectangle has length x cm and width $(x - 5)$ cm.

  a Express the area A cm^2 in terms of x.

Step 1: Use the area formula.

  b Find the rate of change of A with respect to x when $x = 8$.

Step 2: Write terms in the form ax^n.

a $A = x(x - 5)$

Step 3: Differentiate to find $\dfrac{dA}{dx}$.

b $A = x^2 - 5x$

$$\frac{dA}{dx} = 2x - 5$$

Step 4: Substitute the x-value into $\dfrac{dA}{dx}$.

When $x = 8$, $\dfrac{dA}{dx} = 2(8) - 5 = 11$.

The rate of change of A with respect to x when $x = 8$ is 11.

1 Differentiate with respect to x.

 a $y = x^4$ **b** $f(x) = 7x^3 - 2x^2 + 3$ **c** $f(x) = (4x + 1)^2$ **d** $y = \frac{1}{2}x(2x - 1)$

 2 Find the derivative, with respect to x, of

 a $y = x(x - 1)^2$ **b** $y = \dfrac{3x^4 + 5x}{2}$ **c** $y = 2$.

3 For each of the curves in question **2**, find the gradient at $x = 5$.

4 In each of the following, find   **i** $\dfrac{dy}{dx}$ **ii** $\dfrac{d^2y}{dx^2}$

 a $y = x^{-3}$ **b** $y = \dfrac{2}{x^5}$ **c** $y = 4x^{\frac{5}{2}}$ **d** $y = \dfrac{6}{\sqrt{x}}$

5 It is given that $\dfrac{x^2 - x^{\frac{3}{2}}}{2x^2}$.

 a Find $\dfrac{dy}{dx}$.

 b Find the gradient of the curve at $x = 1$.

 6 Evaluate $f'(4)$, where $f(x) = (2x - 3\sqrt{x})^2$. Also find $f''(x)$.

7 The sketch shows the curve $y = \dfrac{3}{2x^2}$.

The points A, B and C on the
curve have x-coordinates -2, -1
and 3 respectively.

Find the gradient of the curve
at each of the points A, B and C.

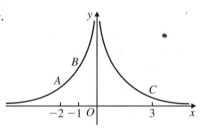

 8 Find the values of x for which the gradient of the curve $y = x^2 + 16x^{-2}$ is zero.

9 Find the values of x for which the gradient of the curve $y = \dfrac{3}{x} + 9x$ is the same as the gradient of the
line $y = 3x + 1$.

10 Find the gradient of the curve $y = 2\sqrt{x^3} - 4x$ at the origin.

11 At time t seconds the length l centimetres of the edge of an expanding square is given by $l = 2\sqrt{t}$.
Find the rate of change of the length of the square at $t = 4$.
Show that the rate of change of the area of the square is constant.

12 The radius r m of an expanding sphere at time t seconds is given by $r = 0.5t^{\frac{2}{3}}$.

 a Find expressions for the surface area, A m², and the volume, V m³, of the sphere at time t seconds.

 b Find the rate of change of the surface area of the sphere at the instant when $t = 27$.

 c Find the rate of change of the volume of the sphere at the instant when $t = 27$.

13 It is given that $V = x(x - 2)^2$. Find the rate of change of V with respect to x when $x = 3$.

SKILLS CHECK **4A EXTRA is on the CD**

4.4 Tangents and normals

Applications of differentiation to tangents and normals.

The **tangent** at P is the line that touches the curve at P.

The **normal** at P is the line through P perpendicular to the tangent.

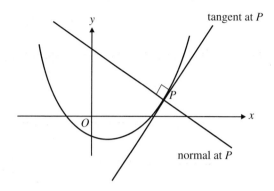

The tangent and normal are perpendicular, so the product of their gradients is -1.

To find the **equation of the tangent** to the curve $y = f(x)$ at the point $P(x_1, y_1)$, find the gradient of the tangent, then substitute into the equation of a line:

$$y - y_1 = m(x - x_1).$$

To find the **equation of the normal**, first find the gradient of the tangent and use it to calculate the gradient of the normal. You can then substitute into the equation of a line, as before.

> **Note:**
> If gradient of tangent $= 2$, then gradient of normal $= -\frac{1}{2}$ (Section 2.2).

> **Recall:**
> Equation of a straight line (Section 2.1)

> **Note:**
> You could use $y = mx + c$.

Example 4.14 A curve has equation $y = x^3 - 6x + 1$. The point P where $x = 2$ lies on the curve. Giving your answers in the form $ax + by + c = 0$, where a, b and c are integers, find the equation of

  a the tangent at P

  b the normal at P.

Step 1: Find $\dfrac{dy}{dx}$.

a $y = x^3 - 6x + 1$

$\dfrac{dy}{dx} = 3x^2 - 6$

Step 2: Substitute x-value into $\dfrac{dy}{dx}$ to give gradient of tangent.

When $x = 2$, $\dfrac{dy}{dx} = 3(2)^2 - 6 = 6$.

The gradient of the tangent when $x = 2$ is 6.

Step 3: Substitute x-value into y to give the y-coordinate.

When $x = 2$, $y = 2^3 - 6(2) + 1 = -3$.

Equation of tangent at P:

Step 4: Use $y - y_1 = m(x - x_1)$.

$$y - (-3) = 6(x - 2)$$
$$y + 3 = 6x - 12$$
$$6x - y - 15 = 0$$

> **Note:**
> $(x_1, y_1) = (2, -3)$ and $m = 6$.

Step 5: Find the gradient of the normal.

b Since product of gradients of perpendicular lines is -1, gradient of normal $= -\frac{1}{6}$.

Equation of normal:

Step 6: Use $y - y_1 = m(x - x_1)$.

$$y - (-3) = -\tfrac{1}{6}(x - 2)$$
$$y + 3 = -\tfrac{1}{6}(x - 2)$$
$$6(y + 3) = -x + 2$$
$$x + 6y + 16 = 0$$

> **Note:**
> $(x_1, y_1) = (2, -3)$ and $m = -\frac{1}{6}$.

Example 4.15 A curve has equation $y = \dfrac{4}{x^3} - \dfrac{x^2}{4}$ and the point P lies on the curve, with x-coordinate 2.

a The equation of the tangent at P is $px + qy = r$, where p, q and r are integers. Find the values of p, q and r.

b Find the equation of the normal at P.

a $y = \dfrac{4}{x^3} - \dfrac{x^2}{4} = 4x^{-3} - \tfrac{1}{4}x^2$

Tip:
Write all terms in the form ax^n before finding $\dfrac{dy}{dx}$.

Step 1: Find $\dfrac{dy}{dx}$.

$$\dfrac{dy}{dx} = -12x^{-4} - \tfrac{1}{2}x$$

Step 2: Substitute the x-value to get the gradient of the tangent at P.

When $x = 2$, $\dfrac{dy}{dx} = -12 \times 2^{-4} - \tfrac{1}{2} \times 2 = -\tfrac{7}{4}$.

The gradient of the tangent at P is $-\tfrac{7}{4}$.

Tip:
Take care with negatives.

Step 3: Substitute the x-value into y to give the y-coordinate.

When $x = 2$, $y = \dfrac{4}{2^3} - \dfrac{(2)^2}{4} = -\tfrac{1}{2}$.

Equation of the tangent at P:

Step 4: Use $y - y_1 = m(x - x_1)$.

$$y - (-\tfrac{1}{2}) = -\tfrac{7}{4}(x - 2)$$
$$y + \tfrac{1}{2} = -\tfrac{7}{4}(x - 2)$$
$$4(y + \tfrac{1}{2}) = -7(x - 2)$$
$$4y + 2 = -7x + 14$$
$$7x + 4y = 12$$

So $p = 7$, $q = 4$, $r = 12$.

Tip:
Rearrange the equation to the given format.

Step 5: Find the gradient of the normal at P.

b At P, gradient of tangent $= -\tfrac{7}{4}$

$\Rightarrow$ gradient of normal $= \tfrac{4}{7}$

Tip:
$m_1 \times m_2 = -1$, so find the negative reciprocal.

Equation of normal at P:

Step 6: Use $y - y_1 = m(x - x_1)$.

$$y - (-\tfrac{1}{2}) = \tfrac{4}{7}(x - 2)$$
$$y + \tfrac{1}{2} = \tfrac{4}{7}(x - 2)$$

Multiply by 14:
$$14y + 7 = 8x - 16$$
$$14y = 8x - 23$$

SKILLS CHECK **4B: Tangents and normals**

1 Find the equation of the tangent to the curve $y = x^3 - 3$ at the point $(1, -2)$.

2 A curve has equation $y = x^3 + 2x - 1$.

a The curve goes through the point $P(1, q)$. Find q.

b Find the gradient at P.

c Find the equation of the tangent at P, giving your answer in the form $y = mx + c$.

d Find the equation of the normal at P, giving your answer in the form $ax + by + c = 0$.

3 A curve has equation $y = 2x^{\frac{3}{2}} - 4x^{\frac{5}{2}} + 2x$.

    a Find the gradient function of the curve.

    b Show that the equation of the tangent to the curve at $(1, 0)$ is $y + 5x = 5$.

    c Find the equation of the normal at the point $(1, 0)$, writing your answer in the form $ax + by + c = 0$, where a, b and c are integers.

4 Find the equation of the normal to the curve $y = \dfrac{4}{x} + x^2$ at the point where $x = 1$, giving your answer in the form $ax + by + c = 0$, where a, b and c are integers.

5 Find the equation of the tangent to the curve $y = 40\sqrt{x}$ at the point where the gradient of the curve is 5, writing your answer in the form $y = mx + c$.

6 A curve has equation $y = \dfrac{x^2 - 2x^3}{x^5}$. The point $P(-1, -3)$ lies on the curve.

    Find the equation of the tangent to the curve at P in the form $y = mx + c$.

 7 It is given that $\mathrm{f}(x) = \dfrac{1}{\sqrt[3]{x}}$.

    a Show that $\mathrm{f}'(8) = -\frac{1}{48}$.

    b Hence find the equation of the tangent to the curve $y = \mathrm{f}(x)$ at the point where $x = 8$, giving your answer in the form $ax + by + c = 0$, where a, b and c are integers.

 8 **a** Sketch the graph of $y = \mathrm{f}(x)$, where $\mathrm{f}(x) = 3x^2 + 1$.

    b Find $\mathrm{f}'(x)$ and use it to find the gradient of the tangent at $x = -2$.
Sketch the tangent on the graph.

9 The sketch shows the curve $y = x^2 - 2x + 3$ and the normal to the curve at $(0, 3)$.

The normal intersects the curve again at A.

    a Find the gradient of the tangent at $(0, 3)$.

    b Find the equation of the normal at $(0, 3)$.

    c Find the x-coordinate of A.

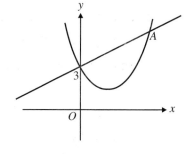

 10 **a** Find the equation of the normal to the curve $y = x^2 + 3$ at the point where $x = -1$.

    b By solving simultaneous equations, find the coordinates of the point where the normal meets the curve again.

    c Draw a sketch to illustrate the information.

SKILLS CHECK **4B EXTRA** is on the CD

1 a Expand $(x^{\frac{3}{2}} - 1)(x^{-\frac{1}{2}} + 1)$.

A curve has equation $y = (x^{\frac{3}{2}} - 1)(x^{-\frac{1}{2}} + 1)$, $x > 0$.

b Find $\dfrac{dy}{dx}$.

c Use your answer to **b** to calculate the gradient of the curve at the point where $x = 4$.

[Edexcel June 2001]

2 A curve C has equation $y = x^3 - 5x^2 + 5x + 2$.

a Find $\dfrac{dy}{dx}$ in terms of x.

The points P and Q lie on C. The gradient of C at both P and Q is 2. The x-coordinate of P is 3.

b Find the x-coordinate of Q.

c Find an equation for the tangent to C at P, giving your answer in the form $y = mx + c$, where m and c are constants.

This tangent intersects the coordinate axes at the points R and S.

d Find the length of RS, giving your answer as a surd.

[Edexcel January 2002]

3 The curve C has equation $y = \dfrac{x^2 + 5x - 3}{3x^{\frac{1}{2}}}$, $x > 0$.

a Find $\dfrac{dy}{dx}$.

b Hence, find the gradient of the normal to the curve at the point where $x = 1$.

4 A curve C has equation given by:

$$y = (2x + 5)(x - 1).$$

a Find $\dfrac{dy}{dx}$.

b Hence, find the equation of the normal to the curve C at the point where $x = 0$ in the form $y = mx + c$.

The normal intersects C again at the point Q.

c Find the exact x-coordinate of the point Q.

5 Curve C has equation $y = x^3 - kx^2$ where k is a constant. The tangent to the curve C at the point with x-coordinate 1 cuts the y-axis at $(0, 2)$. Find:

a the value of k,

b the equation of the tangent in the form $ay + bx + c = 0$.

6 The function $f(x)$ is given by

$$f(x) = \dfrac{(2x - 3)^2}{2\sqrt{x}}, \quad x > 0.$$

a Show that $f(x)$ can be written in the form $Ax^{\frac{3}{2}} + Bx^{\frac{1}{2}} + Cx^{-\frac{1}{2}}$ and state the values of A, B and C.

b Find $f'(x)$.

c Hence, find the gradient of the curve $y = f(x)$ at the point where $x = 1$.

7 A curve has equation $y = \sqrt{x} + x^2 - 4$, $x > 0$. The point A has x-coordinate 1.

    a Find the equation of the normal to the curve at A, leaving your answer in the form $ax + by + c = 0$ where a, b and c are integers.

    b The normal crosses the x-axis at P and the y-axis at Q. Find the coordinates of P and Q.

    c Find the area of the triangle OPQ, where O is the origin.

8 A curve has equation $y = (x - 3)^2 + 4$.

    a Find the equation of the tangent to the curve at the point where $x = 2$, leaving your answer in the form $y = mx + c$.

    b Find the x-coordinate of the point on the curve where the normal to the curve is parallel to the line $y = 2x - 5$.

9 The sketch shows the curve $y = x^2 - 3x + 1$.
The point P has x-coordinate 1 and lies on the curve.

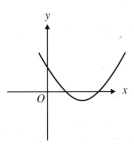

    a Find $\dfrac{dy}{dx}$ when $x = 1$.

    b Hence find the gradient of the normal at P.

    c Find the equation of the normal at P.

    d The normal at P meets the curve again at Q. Find the x-coordinate of Q.

10 The normal at the point $(0, 3)$ on the curve $y = x^2 - 2x + 3$ meets the curve again at A. Find the coordinates of A.

5 Integration

5.1 The reverse of differentiation

Indefinite integration as the reverse of differentiation.

The reverse process of differentiation is called **integration**.

This means that, given $\dfrac{dy}{dx}$, you integrate to find y; given $f'(x)$, you integrate to find $f(x)$.

Indefinite integration

You will have noticed when differentiating that more than one function can have the same derivative, for example:

$$y = 4x^2 \qquad \Rightarrow \frac{dy}{dx} = 8x$$

$$y = 4x^2 - 6 \Rightarrow \frac{dy}{dx} = 8x$$

In fact, for any constant c:

$$y = 4x^2 + c \Rightarrow \frac{dy}{dx} = 8x$$

When integrating, you take this into account by including a constant, called an **integration constant**. All possible values of the constant give a **family** of solutions, known as the **general solution**.

This process is called **indefinite integration**. There are several ways of writing it, including:

$$\frac{dy}{dx} = 8x \Rightarrow y = 4x^2 + c$$

$$f'(x) = 8x \Rightarrow f(x) = 4x^2 + c$$

$$\int 8x\,dx = 4x^2 + c$$

> **Recall:**
> If $y = ax^n$, $\dfrac{dy}{dx} = nax^{n-1}$
> (Section 4.2).

> **Recall:**
> The derivative of a constant is zero (Section 4.2).

> **Note:**
> Often c is used for the integration constant. It can take any value, positive, negative or zero.

> **Note:**
> $\int 8x\,dx$ is read 'the integral of $8x$, with respect to x'.

5.2 Integration methods

Integration of x^n.

To integrate x^n, reverse the differentiation process as follows:

- increase the power by 1
- divide by the new power.

For any rational number n, where $n \neq -1$:

$$\int x^n\,dx = \frac{1}{n+1}x^{n+1} + c$$

> **Note:**
> You can write
> $$\int x^n\,dx = \frac{x^{n+1}}{n+1} + c,\ n \neq -1.$$

For constant function k: $\int k \, dx = kx + c$

If a is a constant:

$$\int a f(x) \, dx = a \int f(x) \, dx \Rightarrow \int ax^n \, dx = \frac{a}{n+1} x^{n+1} + c$$

Note:
You may prefer to write
$\dfrac{ax^{n+1}}{n+1}$.

Example 5.1 Find   **a** $\int x^7 \, dx$ **b** $\int 3 \, dx$ **c** $\int 3x^4 \, dx$.

Step 1: Apply the integration formula.

a $\int x^7 dx = \frac{1}{8} x^8 + c$

b $\int 3 dx = 3x + c$

c $\int 3x^4 dx = \frac{3}{5} x^5 + c$

Example 5.2 Find   **a** $\int 5x\sqrt{x} \, dx$ **b** $\int \sqrt[3]{x} \, dx$.

Step 1: Write term(s) in the form ax^n using the index laws.

Step 2: Integrate and simplify if necessary.

a $\int 5x\sqrt{x} \, dx = \int 5x^{\frac{3}{2}} \, dx$

$= \frac{5}{\frac{5}{2}} x^{\frac{5}{2}} + c$

$= 2x^{\frac{5}{2}} + c$

b $\int \sqrt[3]{x} \, dx = \int x^{\frac{1}{3}} \, dx$

$= \frac{1}{\frac{4}{3}} x^{\frac{4}{3}} + c$

$= \frac{3}{4} x^{\frac{4}{3}} + c$

Recall:
Laws of indices (Section 1.1).

Tip:
Take special care with fractions.

Tip:
When the term contains fractions it may be easier to use
$\int ax^n dx = a \int x^n dx$
$= a \times \dfrac{1}{n+1} x^{n+1} + c.$

To integrate an expression in x containing several terms, such as $x^3 + 5x^2 + 2x - 1$, integrate term by term, using the rule

$$\int (f(x) \pm g(x)) \, dx = \int f(x) \, dx \pm \int g(x) \, dx.$$

Example 5.3 Find $\int (x^3 + 5x) \, dx$.

Step 1: Integrate term by term.

$\int (x^3 + 5x) \, dx = \frac{1}{4} x^4 + \frac{5}{2} x^2 + c$

Tip:
Do not forget to include the integration constant c.

If you are given $\dfrac{dy}{dx}$, the gradient function, then to find y, integrate $\dfrac{dy}{dx}$:

$$y = \int \frac{dy}{dx} \, dx + c.$$

Example 5.4 Given that $\dfrac{dy}{dx} = 5x^4 - 6x + 3$, express y in terms of x.

Step 1: Integrate term by term.

Step 2: Simplify where possible.

$y = \int (5x^4 - 6x + 3) \, dx$

$= \frac{5}{5} x^5 - \frac{6}{2} x^2 + 3x + c$

$= x^5 - 3x^2 + 3x + c$

Tip:
This is asking you to integrate the function of x.

If you are given $f'(x)$ then integrate to find $f(x)$:

$$f(x) = \int f'(x)\,dx + c.$$

Example 5.5 Given that $f'(x) = x^2(3 - 2x^4)$, find $f(x)$.

Step 1: Write term(s) in the form ax^n.

$$f'(x) = x^2(3 - 2x^4)$$
$$= 3x^2 - 2x^6$$

Step 2: Integrate term by term.

$$f(x) = \int (3x^2 - 2x^6)\,dx$$
$$= \tfrac{3}{3}x^3 - \tfrac{2}{7}x^7 + c$$

Step 3: Simplify where possible.

$$= x^3 - \tfrac{2}{7}x^7 + c$$

Tip:
Expand the bracket.

Sometimes you have additional information enabling you to find a specific value for the integration constant. In this case you can give a **particular solution**, rather than a general one, as in Example 5.6.

Example 5.6 A curve with gradient function $2x(1 - 3x)$ goes through the point $P(2, -10)$. Find the equation of the curve.

Step 1: Write term(s) in the form ax^n.

$$\frac{dy}{dx} = 2x(1 - 3x) = 2x - 6x^2$$

Step 2: Integrate term by term to find y.

$$y = \int (2x - 6x^2)\,dx = x^2 - 2x^3 + c$$

Step 3: Substitute the coordinates of the given point to find c.

At P, $x = 2$ and $y = -10$.

Substituting into $y = x^2 - 2x^3 + c$ gives

$$-10 = 2^2 - 2 \times 2^3 + c$$
$$-10 = 4 - 16 + c$$
$$c = 2$$

The equation of the curve is $y = x^2 - 2x^3 + 2$.

Recall:
The gradient function is $\dfrac{dy}{dx}$
and $y = \displaystyle\int \frac{dy}{dx}\,dx + c$.

Note:
The general solution contains the integration constant c.

Note:
Use the fact that P lies on the curve to find the value of c.

Note:
This is the particular solution.

Example 5.7 Find $\displaystyle\int \left(x + \frac{1}{x}\right)^2 dx$.

Step 1: Write term(s) in the form ax^n using the index laws.

$$\left(x + \frac{1}{x}\right)^2 = \left(x + \frac{1}{x}\right)\left(x + \frac{1}{x}\right)$$
$$= x^2 + 1 + 1 + \frac{1}{x^2}$$
$$= x^2 + 2 + x^{-2}$$

Step 2: Integrate term by term.

$$\int \left(x + \frac{1}{x}\right)^2 dx = \int (x^2 + 2 + x^{-2})\,dx$$
$$= \frac{1}{3}x^3 + 2x + \frac{1}{-1}x^{-1} + c$$
$$= \tfrac{1}{3}x^3 + 2x - x^{-1} + c$$

Tip:
Expand the brackets.

Recall:
$\dfrac{1}{x^a} = x^{-a}$ (Section 1.1).

Tip:
You could write this as
$\dfrac{1}{3}x^3 + 2x - \dfrac{1}{x} + c$

Example 5.8 **a** Write $\dfrac{4x^6 - x}{2x^4}$ in the form $ax^p + bx^q$ where a, b, p and q are rational numbers to be found.

b Hence find $\displaystyle\int \frac{4x^6 - x}{2x^4}\,dx$.

Tip:
This is telling you to write the expression in index form before integrating.

Step 1: Write term(s) in the form ax^n using the index laws.

a $\dfrac{4x^6 - x}{2x^4} = \dfrac{4x^6}{2x^4} - \dfrac{x}{2x^4} = 2x^2 - \tfrac{1}{2}x^{-3}$

Comparing coefficients gives $a = 2$, $b = -\tfrac{1}{2}$, $p = 2$, $q = -3$.

Recall:
$\dfrac{x^m}{x^n} = x^{m-n}$ (Section 1.1).

Step 2: Integrate term by term.

b $\displaystyle\int \dfrac{4x^6 - x}{2x^4}\,dx = \int (2x^2 - \tfrac{1}{2}x^{-3})\,dx$

$$= \dfrac{2}{3}x^3 - \dfrac{\frac{1}{2}}{-2}x^{-2} + c$$

$$= \tfrac{2}{3}x^3 + \tfrac{1}{4}x^{-2} + c$$

Tip:
Use your answer to part **a**.

Tip:
Be very careful with negatives and fractions. It is a good idea to write down the working.

Example 5.9 A curve goes through the point $(1, 4)$ and has gradient function $\dfrac{4}{x^3}$. Find the equation of the curve.

Step 1: Write term(s) in the form ax^n using the index laws.

$\dfrac{dy}{dx} = \dfrac{4}{x^3} = 4x^{-3}$

Tip:
Write the gradient function as $\dfrac{dy}{dx}$.

Step 2: Integrate term by term.

$y = \displaystyle\int 4x^{-3}\,dx$

$$= \dfrac{4}{-2}x^{-2} + c$$

$$= -2x^{-2} + c$$

Tip:
Remember to include the integration constant.

Step 3: Substitute the coordinates of the given point to find c.

Since $(1, 4)$ lies on the curve, when $x = 1$, $y = 4$.

Substituting into the equation:

$4 = -2(1)^{-2} + c \Rightarrow c = 6$

The equation of the curve is $y = -2x^{-2} + 6$.

Tip:
You could write $y = 6 - \dfrac{2}{x^2}$.

Example 5.10 Given that $f'(x) = \dfrac{(x+2)^2}{3\sqrt{x}}$ and that $f(0) = 1$, find $f(x)$.

Step 1: Write term(s) in the form ax^n using the index laws.

$f'(x) = \dfrac{(x+2)^2}{3\sqrt{x}} = \dfrac{x^2 + 4x + 4}{3x^{\frac{1}{2}}}$

Tip:
Expand the numerator and then divide each term in the numerator by $3x^{\frac{1}{2}}$.

$$= \dfrac{x^2}{3x^{\frac{1}{2}}} + \dfrac{4x}{3x^{\frac{1}{2}}} + \dfrac{4}{3x^{\frac{1}{2}}}$$

$$= \tfrac{1}{3}x^{\frac{3}{2}} + \tfrac{4}{3}x^{\frac{1}{2}} + \tfrac{4}{3}x^{-\frac{1}{2}}$$

Step 2: Integrate term by term.

$f(x) = \displaystyle\int (\tfrac{1}{3}x^{\frac{3}{2}} + \tfrac{4}{3}x^{\frac{1}{2}} + \tfrac{4}{3}x^{-\frac{1}{2}})\,dx$

Tip:
$f(x) = \displaystyle\int f'(x)\,dx + c$

$$= \dfrac{\frac{1}{3}}{\frac{5}{2}}x^{\frac{5}{2}} + \dfrac{\frac{4}{3}}{\frac{3}{2}}x^{\frac{3}{2}} + \dfrac{\frac{4}{3}}{\frac{1}{2}}x^{\frac{1}{2}} + c$$

$$= \tfrac{2}{15}x^{\frac{5}{2}} + \tfrac{8}{9}x^{\frac{3}{2}} + \tfrac{8}{3}x^{\frac{1}{2}} + c$$

Recall:
$\dfrac{\frac{1}{3}}{\frac{5}{2}} = \tfrac{1}{3} \div \tfrac{5}{2} = \tfrac{1}{3} \times \tfrac{2}{5}$

Substituting $x = 0$: $f(0) = 0 + 0 + 0 + c = c$

Step 3: Substitute the coordinates of the given point to find c.

$f(0) = 1 \Rightarrow c = 1$

$f(x) = \tfrac{2}{15}x^{\frac{5}{2}} + \tfrac{8}{9}x^{\frac{3}{2}} + \tfrac{8}{3}x^{\frac{1}{2}} + 1$

1 Find

a $\int 4x^2 dx$ 　　　　 **b** $\int \frac{2}{5}x^2 dx$ 　　　　 **c** $\int -6 dx$

2 Find y in terms of x if:

a $\dfrac{dy}{dx} = x^5 + 3x^2$ 　　 **b** $\dfrac{dy}{dx} = 2x^4 - \frac{1}{2}x^3$ 　　 **c** $\dfrac{dy}{dx} = x(2x + 7)$

d $\dfrac{dy}{dx} = 4 - 3x$ 　　 **e** $\dfrac{dy}{dx} = -\frac{2}{5}x^3 - 1$ 　　 **f** $\dfrac{dy}{dx} = 2c$

 3 a Expand $(2x + 3)(x - 4)$. 　　　 **b** Find $\int (2x + 3)(x - 4) dx$.

4 a Given that $\dfrac{dA}{dt} = 4t^7$, find A in terms of t.

　　b Given that $A = 0$ when $t = 1$, find the value of A when $t = 2$.

5 A curve passes through the origin and has gradient function $5x - 3$. Find the equation of the curve.

6 Find

a $\int \dfrac{3}{x^2} dx$ 　　　　 **b** $\int \dfrac{3}{\sqrt{x}} dx$ 　　　　 **c** $\int \sqrt[4]{x}\, dx$

7 a Express $x^2\sqrt{x}$ in the form x^k, where k is a rational number. 　　 **b** Find $\int x^2\sqrt{x}\, dx$.

8 Find 　　 **a** $\int (x^2 + \sqrt{x})\, dx$ 　　　 **b** $\dfrac{x + 2}{\sqrt{x}} dx$.

 9 A curve passes through the point $(1, 3)$ and has gradient function $\dfrac{1}{2\sqrt{x}}$.
Find the equation of the curve.

Examination practice 　Integration

1 Find $\int \dfrac{2}{x^2} dx$.

2 $y = 7 + 10x^{\frac{3}{2}}$.

a Find $\dfrac{dy}{dx}$.

b Find $\int y\, dx$.

[Edexcel January 2003]

3 The curve C with equation $y = f(x)$ is such that

$$\frac{dy}{dx} = 3\sqrt{x} + \frac{12}{\sqrt{x}}, \ x > 0.$$

   a Show that, when $x = 8$, the exact value of $\frac{dy}{dx}$ is $9\sqrt{2}$.

   The curve C passes through the point $(4, 30)$.

   b Using integration, find $f(x)$.   [Edexcel June 2004]

4 The curve C has equation $y = f(x)$. Given that

$$\frac{dy}{dx} = 3x^2 - 20x + 29$$

   and that C passes through the point $P(2, 6)$,

   a find y in terms of x.

   b Verify that C passes through the point $(4, 0)$.

   c Find an equation of the tangent to C at P.

   The tangent to C at the point Q is parallel to the tangent at P.

   d Calculate the exact x-coordinate of Q.   [Edexcel November 2002]

5 $\frac{dy}{dx} = 5 + \frac{1}{x^2}$.

   a Use integration to find y in terms of x.

   b Given that $y = 7$ when $x = 1$, find the value of y at $x = 2$.   [Edexcel June 2003]

 6 The curve C has gradient function given by

$$\frac{dy}{dx} = x^2(7\sqrt{x} - 3), \ x > 0.$$

   a Use integration to find y in terms of x.

   b Given that $y = 5$ at $x = 1$, find y in terms of x.

7 **a** Expand $(2\sqrt{x} - 3)(2\sqrt{x} + 3)$.

   b Hence, find $\int (2\sqrt{x} - 3)(2\sqrt{x} + 3) \, dx$.

8 A curve passes through the point with coordinates $(4, 10)$ and has gradient at (x, y) given by

$$\frac{dy}{dx} = \frac{3}{2}\sqrt{x} - \frac{2}{\sqrt{x^3}}, \ x > 0.$$

   a Find the equation of the curve.

   b Show that the curve also passes through the point with coordinates $(16, 65)$.

9 $f'(x) = k\sqrt{x} - 3x^2 + 41, \ x > 0.$

   a Given that $f'(4) = -1$, find the value of the constant k.

   b Given also that $f(4) = 134$, find $f(x)$.

   c Using the results in parts **a** and **b**, find the equation of the tangent to the curve $y = f(x)$ at the point where $x = 4$, leaving your answer in the form $y = mx + c$.

10 The gradient of a curve at the point (x, y) is given by $\frac{dy}{dx} = 3x^2 - x$. Use integration to find the equation of the curve, given that the curve passes through the point $(2, 1)$.

Practice exam paper

Answer **all** questions.

Time allowed: 1 hour 30 minutes

Calculators may **not** be used in this examination.

1 Find $\int (x^4 - 2x^2 + 1)\,dx$. *(3 marks)*

2 The nth term of a sequence is $17 - 2n$.

    a Write down the first three terms of the sequence. *(2 marks)*

    b Calculate $\displaystyle\sum_{n=1}^{20} (17 - 2n)$. *(2 marks)*

3 Express in the form $a + b\sqrt{2}$, where a and b are integers,

    a $\sqrt{8}(7 - \sqrt{2})$ *(3 marks)*

    b $\dfrac{21}{3 + \sqrt{2}}$. *(3 marks)*

4 The points A and B, with coordinates $(1, 3)$ and $(3, -1)$ respectively, lie on the line with equation $y = -2x + 5$. The midpoint of AB is M.

    a Find the coordinates of M. *(2 marks)*

    b Find an equation of the line perpendicular to AB which passes through M. *(3 marks)*

    c Hence show that the perpendicular bisector of AB passes through the origin O. *(2 marks)*

5

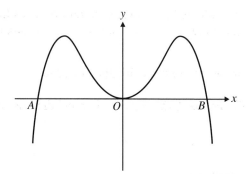

The diagram shows a sketch of a curve with equation $y = f(x)$. The curve touches the x-axis at the origin O and crosses the x-axis at the points A and B with coordinates $(-2, 0)$ and $(2, 0)$ respectively.

On separate diagrams sketch the curve with equation

    a $y = f(x + 2)$ *(3 marks)*

    b $y = f(\tfrac{1}{2}x)$. *(3 marks)*

On each diagram show clearly the coordinates of the points at which the curve crosses or touches the x-axis.

6 **a** Find the set of values of x for which $4(x - 2) > 7 - 2(x - 3)$. *(3 marks)*

    b The equation $kx^2 + kx + 2 = 0$, where k is a constant, has no real roots. Find the set of possible values of k. *(4 marks)*

7 A curve C has equation $y = lx^3 + mx + 2$, where l and m are constants. Given that the point P with coordinates $(-1, 4)$ lies on C,

    a find an equation satisfied by l and m. *(2 marks)*

Given also that the gradient of the normal to C at P is parallel to the line with equation $4y + x = 8$,

    b find a second equation satisfied by l and m. *(5 marks)*

    c Hence show that $l = 3$ and find the value of m. *(3 marks)*

8 **a** Solve the equation $(x + 2)^2 = 4$. *(2 marks)*

A curve C has equation $y = (x + 2)^2 - 4$.

    b Sketch the curve C, showing clearly the coordinates of any points of intersection with the coordinate axes. *(2 marks)*

    c Show that the x-coordinates of the points of intersection of C with the line l with equation $y = 2x + 3$ satisfy the equation $x^2 + 2x - 3 = 0$. *(3 marks)*

    d Hence calculate the coordinates of the points of intersection of C and l. *(3 marks)*

9 A student is answering a test in which questions are presented one at a time on a computer screen. The student is allowed 20 seconds to answer the first question. For each subsequent question, the student is allowed 5 seconds more to answer the question than was allowed for the previous question. Hence the student is allowed 25 seconds to answer the second question and 30 seconds to answer the third question.

    a Calculate the time allowed to answer the tenth question. *(2 marks)*

A test consists of 20 questions.

    b Calculate the total time allowed for answering all 20 questions in this test. *(3 marks)*

A second test is set in which the time allowed to answer each question has the same conditions as those specified above. The total time allowed to answer all of the questions in this test is 2000 seconds.

    c Use algebra to calculate the number of questions in this second test. *(5 marks)*

10 A curve C has equation $y = f(x)$. Given that $f'(x) = 3x^2 - 6\sqrt{x}$ and that the point P with coordinates $(1, -5)$ lies on C,

    a find $f(x)$, *(6 marks)*

The point Q lies on C and has x-coordinate 4.

    b Show that an equation of the tangent to C at Q is $y = 36x - 114$. *(6 marks)*

Answers

1 **a** $x = 4, y = 1$ **b** $x = 3, y = 2$ **c** $a = 3, b = -2$
2 **a** $x = -6, y = 4$ **b** $a = 11, b = 3$ **c** $p = -9, q = 3$
3 **b** $x = 4, y = -\frac{2}{3}$
4 **a** $x = 2, y = 8$ or $x = -6, y = 16$
   b $x = \frac{2}{3}, y = 2$ or $x = -1, y = -3$
   c $x = 0, y = 1$ or $x = -1, y = 0$
5 width $= 3\,\text{cm}$, length $= 6\,\text{cm}$
6 **b** $2a + b = 4$ **c** $a = 3, b = -2$

SKILLS CHECK 1A (page 3)

1 **a** $6x^4 y^4$ **b** $16a^8$ **c** $7p^{-1}q^2$
2 **a** 2 **b** $\frac{1}{9}$ **c** $\frac{1}{3}$ **d** 64 **e** 2
3 $12a^5 b^{-\frac{3}{2}}$
4 **a** $x^{\frac{5}{2}}$ **b** x^{-1} **c** x^5
5 $p^{\frac{1}{3}}$
6 **a** **i** 2^{2x} **ii** 2^{3x-3} **b** 2^{5x-3} **c** 3
7 **a** 3^{4x} **b** $-\frac{1}{3}$ **c** $-\frac{3}{4}$
8 **a** $y = 3x - \frac{3}{2}$ **b** $4y = 2x + 9$ **c** $x = \frac{3}{2}, y = 3$
9 $x = 2$
10 **a** 6^{4p} **b** 6^{3q-3} **c** $4p = 3q - 3$ **d** $p = -1, q = -\frac{1}{3}$

SKILLS CHECK 1F (page 19)

1 **a** $x > 3$ **b** $x > -1$ **c** $x \geqslant 8\frac{2}{3}$ **d** $y > 1$
   e $x > 6$ **f** $x \leqslant 14.5$
2 **a** $3 < x < 9$ **b** $-6 \leqslant x \leqslant 10$
3 $a = 3$
4 **a** $y < -2, y > 2$ **b** $-7 \leqslant x \leqslant 7$ **c** $x \leqslant -\sqrt{5}, x \geqslant \sqrt{5}$
   d $-3 < x < 3$ **e** $x < -1, x > 3$
   f $-2 - \sqrt{5} \leqslant x \leqslant -2 + \sqrt{5}$
5 **a** $p = 2, q = -9$ **b** $x \leqslant -5, x \geqslant 1$
6 **a** $-4 < x < 3$ **b** $x \leqslant -2\frac{1}{2}, x \geqslant \frac{2}{3}$ **c** $x \leqslant -5, x \geqslant 4$
   d $-5 < p < -2$ **e** $x \leqslant -2, x \geqslant \frac{3}{2}$
7 **a** $p = 2, q = 10$ **c** $2 - \sqrt{10} < x < 2 + \sqrt{10}$
8 **a** $-6 < k < 6$ **b** $-4 \leqslant k \leqslant 4$
9 **a** $k^2 - 16$ **b** $k < -4, k > 4$

SKILLS CHECK 1B (page 6)

1 **a** $5\sqrt{2}$ **b** $4\sqrt{2}$ **c** $7\sqrt{2}$ **d** $7\sqrt{3}$ **e** $7\sqrt{7} - 4\sqrt{5}$
2 **a** $11\sqrt{3}$ **b** $2\sqrt{2}$ **c** $5 - 2\sqrt{6}$ **d** 50 **e** $2\sqrt{2}$
   f 3 **g** $\frac{\sqrt{3}}{3}$ **h** $\sqrt{6} - 2$
3 **a** $12 + 2\sqrt{11}$ **b** 10 **c** $\frac{6}{5} + \frac{\sqrt{11}}{5}$
4 $p = \frac{3}{2}$
5 $4\sqrt{3}, 2\sqrt{3}$ **b** $6\sqrt{3}$
6 **a** $8 + 2\sqrt{7}$ **b** $\frac{4}{3} + \frac{1}{3}\sqrt{7}$
7 **a** $p = 7, q = 1$ **b** $r = \frac{7}{11}, s = \frac{1}{11}$

SKILLS CHECK 1G (page 21)

1 **a** $-x^2 - 18x + 5$ **b** $4x^3 + 14x^2 - 7x + 4$
   c $x^3 - 5x^2 - 9x + 45$ **d** $2x^3 + 3x^2 - 23x - 12$
2 **a** $a = 2, b = 5$ **b** $a = 3, b = -1, c = 5$
   c $a = -4, b = -\frac{3}{2}, c = -2$ or $a = 3, b = 2, c = -2$
3 **a** $x(x-2)(x+5)$ **b** $2x(x+1)^2$ **c** $x(x-3)(x+3)$
   d $x(1-x)(6+x)$ **e** $x(x+3)(x-7)$

SKILLS CHECK 1C (page 10)

1 **a** $x(x+5)$ **b** $(x-1)(x-1)$ **c** $(a+4)(a-4)$
   d $(x+1)(x-6)$ **e** $(x+15)(x-2)$ **f** $2x(x-4)$
2 **a** $(2x+3)(x+2)$ **b** $(5x+1)(x-3)$ **c** $(3y-2)(y+2)$
   d $4(1-2x)(5x+3)$ **e** $(2x+5)(2x-5)$
3 **a** $(x+3)^2 - 1$ **b** $(x-6)^2 - 39$ **c** $\left(x + \frac{5}{2}\right)^2 - \frac{33}{4}$
4 **a** $2\left(x + \frac{7}{4}\right)^2 - \frac{1}{8}$ **b** $5\left(x - \frac{7}{5}\right)^2 - \frac{64}{5}$ **d** $3\left(y + \frac{2}{3}\right)^2 - \frac{16}{3}$
   e $-40\left(x + \frac{1}{20}\right)^2 + \frac{121}{10}$ **f** $4x^2 - 25$
5 **a** $p = 2, q = 3$ **b** $(2, 3)$ **c** Minimum   **d** $x = 2$
6 **a** $4(x+1)^2 - 3$ **b** $(-1, -3)$, minimum   **c** $x = -1$
7 **a** $18 - (x+2)^2$ **b** 18 **c** $(-2, 18)$

SKILLS CHECK 1H (page 26)

2 **a** $(1-2x)(x+4)$
4 **a** $(x+6)(x-4), x = -6, x = 4$ **b** $(x+1)^2 - 25$ **d** $x = -1$
7 **a** $x = 1, y = 0$ **b** $x = 2, y = 8$

SKILLS CHECK 1I (page 32)

1 **a** **i** $\begin{pmatrix} 0 \\ 3 \end{pmatrix}$ **ii** $(0, 3), 3$ **b** **i** $\begin{pmatrix} 0 \\ -2 \end{pmatrix}$ **ii** $(0, -2), -2$
   c **i** $\begin{pmatrix} 0 \\ 1 \end{pmatrix}$ **ii** $(0, 1), 1$
2 **a** $\begin{pmatrix} -2 \\ 0 \end{pmatrix}$ **b** $\begin{pmatrix} 1 \\ 0 \end{pmatrix}$ **c** $\begin{pmatrix} -4 \\ 0 \end{pmatrix}$
3 **a** $a = 1, b = -1$ **b** $a = -3, b = 2$
4 **a** $y = (x-3)^2 + 1$ **b** $\begin{pmatrix} 3 \\ 1 \end{pmatrix}, (3, 1)$
   c Graph doesn't cross x-axis.
5 **a** $(1, -3)$
   b **i** Stretch in the y-direction, factor 2
     ii Curve goes through $(0, 0), (1, 6), (2, 0)$; $A(1, 6)$
6 **a** **i** Stretch in the y-direction, factor 4
     ii Stretch in the x-direction, factor $\frac{1}{2}$
   b $y = 4x^2$; the curves are the same
7 **a** Graph through $(-2, 0), (-1, -0.5), (0, 0), (2, 1)$
   b Graph through $(-2, -1), (0, 0), (1, 0.5), (2, 0)$
   c Graph through $(-2, 0), (0, -1), (1, -1.5), (2, -1)$

SKILLS CHECK 1D (page 14)

1 **a** $x = 2, x = -3$ **b** $x = \frac{1}{4}, x = -\frac{3}{2}$ **c** $x = 0, x = -5$
2 **a** $x = -5, x = -1$ **b** $x = 8, x = 3$ **c** $x = 0, x = 6$
   d $x = -1, x = 6$ **e** $x = -2, x = 3$ **f** $x = \pm 6$
3 **a** $\left(x + \frac{3}{2}\right)^2 - \frac{29}{4}$ **b** $x = \frac{-3 \pm \sqrt{29}}{2}$
4 **a** $2\left(x - \frac{3}{4}\right)^2 - \frac{25}{8}$ **b** $x = -\frac{1}{2}, x = 2$
5 $x = \frac{-3 \pm \sqrt{69}}{10}$
6 **a** $x = -1, x = 4$ **b** $x = \frac{7 \pm \sqrt{13}}{4}$
7 $x = -\frac{1}{3} \pm \frac{1}{3}\sqrt{13}$
8 **a** 2 **b** 2 **c** 1 (repeated)   **d** 2
9 -56

Exam practice 1 (page 33)

1 **a** $y = \frac{9}{2}x + \frac{3}{2}$ **b** $x = -\frac{1}{9}$
2 **a** $y = \frac{4}{3}x + \frac{2}{3}$ **b** $x = -\frac{5}{3}, y = -\frac{14}{9}$
3 **a** $a = 1, b = 2$ **b** $c = \frac{1}{9}, d = \frac{2}{9}$

4   **a** $1 + \sqrt{13}, 1 - \sqrt{13}$ **b** $x > 1 + \sqrt{13}$ or $x < 1 - \sqrt{13}$

5 $y = 1$ or -1.6; $(3, 1)$, $(-2.2, -1.6)$

6 $x = -4$, $y = 3\frac{1}{2}$

7 $k = \pm 9$

8   **a** $3, 2, -7$ **b** -7 **c** $-0.5, -3.5$

9   **a** $-5, 4$ **b** $x > 4$ or $x < -5$

SKILLS CHECK 2A (page 38)

1   **a** $3x - y - 2 = 0$ **b** $3x + 2y - 6 = 0$
       c $2x + 3y - 20 = 0$ **d** $6x + 5y - 20 = 0$

2   **a** $y = \frac{4}{5}x - \frac{8}{5}$ **b**   Gradient $= -\frac{2}{3}$, y-intercept $= 2$

3   **a i** -1 **ii** $(\frac{1}{2}, \frac{5}{2})$ **iii** $7\sqrt{2}$
       b i $\frac{5}{3}$ **ii** $(1, 1)$ **iii** $2\sqrt{34}$
       c i $\frac{1}{3}$ **ii** $(\frac{1}{2}, -\frac{3}{2})$ **iii** $\sqrt{10}$

4   **a** $y = \frac{3}{5}x + \frac{2}{5}$ **b** $y = -3x + 12$ **c** $y = -x + 3$

6   **b** $(-3, -1)$ **c**   15 units^2

SKILLS CHECK 2B (page 40)

1   **a**   Perpendicular   **b**   Parallel   **c**   Neither

3 $y = \frac{2}{3}x + 3$

4 $y = -\frac{5}{4}x - 2$

5 $y = 5x - 16$

6 $y = 3x - 7$

7 $5x + 3y + 19 = 0$

8 $x + 5y - 12 = 0$

9   **a** $\dfrac{5}{2}$ **b** $y = \dfrac{5}{2}x - \dfrac{29}{2}$ **c** $-\dfrac{1}{2}$ **d** $y = -\dfrac{x}{2} + \dfrac{1}{2}$ **e** $(5, -2)$

Exam practice 2 (page 41)

1   **a** $(2.5, -2.5)$ **b** $2x + 3y = 1$ **c** $(1\frac{1}{13}, -\frac{5}{13})$

2   **a** $7x + 5y - 18 = 0$ **b** $\frac{162}{35}$ units^2 (accept 4.63 units^2)

3   **a** $y + 2x = 20$ **b** $3y = x + 4$ **c** $4\sqrt{5}$

4   **a** l_1 meets x-axis at $(0, 0)$ and meets y-axis at $(0, 0)$.
          l_2 meets x-axis at $(1.5, 0)$ and meets y-axis at $(0, -3)$.
       b $x = \frac{4}{3}$, $y = -\frac{1}{3}$ **c** $12x - 3y - 17 = 0$

5   **a** $\frac{1}{2}$ **b**   6   **c** $4\sqrt{5}$
       d   10   **e** $2x + y - 16 = 0$ **f** $(4, 8)$

6   **a** $x + 2y = 16$ **b** $y = -4x$ **c** $(\frac{6}{7}, \frac{53}{7})$

7   **a** $p = 2$, $q = 3$ **b** $(1, \frac{3}{2})$ **c** $y = \frac{2}{3}x + \frac{5}{6}$

8   **a i** -3 **ii** $3x + y = 4$ **b i** $(2, -2)$ **ii** $\frac{1}{3}$ **iii** $p = -1$

SKILLS CHECK 3A (page 45)

1 $0, 3, 8, 15$

2 $\frac{1}{3}, \frac{5}{9}, \frac{19}{27}, \frac{65}{81}$

3   **a** $4, 8, 16, 32$ **b** $7, 15, 31, 63$

4 $-9, -4\frac{5}{9}$

5   **a**   75   **b**   31

6   **a**   42   **b**   15

7 $-4, -9, -7$

8   **a**   40   **b** $2\frac{103}{210}$

SKILLS CHECK 3B (page 50)

1   **a** $5, 1190$ **b** $-3, -610$

2 $u_n = n - \frac{1}{2}$, $S_{200} = 20\,000$

3   **a**   3   **b** $3n + 5$ **c**   1100

4   **a**   3   **b** $a = 9$, $d = -1$ **c** -165

5   **a**   £50   **b**   £10 500

6   **a**   741   **b**   1179

7   99

8   9

9   **a** $8, 10, 12$ **b**   2   **c**   20   **d**   540

10 **a**   15 months   **b**   £725

Exam practice 3 (page 51)

1   **a** $25, -3$ **b** -3810

2   **a** $77, 74$ **b** -3 **c**   175

3   **a**   5   **b**   45

4   **a** $2k - 3, 2k^2 - 3k - 3$ **b** $4, -\frac{5}{2}$ **c**   65

5   **a** $d = 5$ **b**   59

6   **b** $\frac{11}{3}k - 3$ **c** $\frac{3}{2}$ **d**   415

7   3 cm

8   **a** $-6, 2$ **b**   260

9   5

10 **a** $4 + k, 16 + 5k$ **b** -2 **c** $22, 86$

SKILLS CHECK 4A (page 58)

1   **a** $4x^3$ **b** $21x^2 - 4x$ **c** $32x + 8$ **d** $2x - \frac{1}{2}$

2   **a** $3x^2 - 4x + 1$ **b** $6x^3 + \frac{5}{2}$ **c**   0

3   **a**   56   **b** $722\frac{1}{2}$ **c**   0

4   **a i** $-3x^{-4}$ **ii** $12x^{-5}$ **b i** $-10x^{-6}$ **ii** $60x^{-7}$
       c i $10x^{\frac{3}{2}}$ **ii** $15x^{\frac{1}{2}}$ **d i** $-3x^{-\frac{3}{2}}$ **ii** $\frac{9}{2}x^{-\frac{5}{2}}$

5 $\frac{1}{4}x^{-\frac{3}{2}}, \frac{1}{4}$

6 $5; 8 - 9x^{-\frac{1}{2}}$

7 $\frac{3}{8}, 3, -\frac{1}{9}$

8 ± 2

9 $\pm \frac{1}{2}\sqrt{2}$

10 -4

11 $0.5 \, \text{cm s}^{-1}$

12 **a** $A = \pi t^{\frac{4}{3}}$, $V = \frac{1}{6}\pi t^2$ **b** $4\pi \, \text{m}^2 \, \text{s}^{-1}$ **c** $9\pi \, \text{m}^3 \, \text{s}^{-1}$

13   7

SKILLS CHECK 4B (page 60)

1 $y = 3x - 5$

2   **a**   2   **b**   5   **c** $y = 5x - 3$ **d** $x + 5y - 11 = 0$

3   **a** $3x^{\frac{1}{2}} - 10x^{\frac{3}{2}} + 2$ **c** $x - 5y - 1 = 0$

4 $x - 2y + 9 = 0$

5 $y = 5x + 80$

6 $y = -7x - 10$

7   **b** $x + 48y - 32 = 0$

8   **b** -12

9   **a** -2 **b** $y = \frac{1}{2}x + 3$ **c** $\frac{5}{2}$

10 **a** $2y = x + 9$ **b** $(\frac{3}{2}, \frac{21}{4})$

Exam practice 4 (page 62)

1   **a** $x + x^{-\frac{1}{2}} + x^{\frac{3}{2}} - 1$ **b** $1 + \frac{1}{2}x^{-\frac{3}{2}} + \frac{3}{2}x^{\frac{1}{2}}$ **c** $4\frac{1}{16}$

2   **a** $\dfrac{dy}{dx} = 3x^2 - 10x + 5$ **b** $\dfrac{1}{3}$
       c $y = 2x - 7$ **d** $\dfrac{7\sqrt{5}}{2}$

3   **a** $\dfrac{dy}{dx} = \dfrac{1}{2}x^{\frac{1}{2}} + \dfrac{5}{6}x^{-\frac{1}{2}} + \dfrac{1}{2}x^{-\frac{3}{2}}$ **b** $-\dfrac{6}{11}$

4   **a** $\dfrac{dy}{dx} = 4x + 3$ **b** $y = -\dfrac{1}{3}x - 5$ **c** $x = -\dfrac{5}{3}$

5   **a** $k = 4$ **b** $y + 5x - 2 = 0$

6   **a** $A = 2, B = -6, C = \frac{9}{2}$ **b** $3x^{\frac{1}{2}} - 3x^{-\frac{1}{2}} - \frac{9}{4}x^{-\frac{3}{2}}$ **c** $-\frac{9}{4}$

7   **a** $2x + 5y + 8 = 0$ **b** $P(4, 0), Q(0, \frac{8}{5})$ **c** $\frac{16}{5}$

8   **a** $y = -2x + 9$ **b** $\frac{11}{4}$

9   **a** -1 **b**   1   **c** $y = x - 2$ **d**   3

10 $A(\frac{5}{2}, \frac{17}{4})$

1 **a** $\dfrac{4}{3}x^3 + c$ **b** $\dfrac{2}{15}x^3 + c$ **c** $-6x + c$

2 **a** $y = \dfrac{1}{6}x^6 + x^3 + c$ **b** $y = \dfrac{2}{5}x^5 - \dfrac{1}{8}x^4 + c$ **c** $y = \dfrac{2}{3}x^3 + \dfrac{7}{2}x^2 + c$

    d $y = -\dfrac{3}{2}x^2 + 4x + c$ **e** $y = -\dfrac{1}{10}x^4 - x + c$ **f** $y = 2cx + k$

3 **a** $2x^2 - 5x - 12$ **b** $\dfrac{2}{3}x^3 - \dfrac{5}{2}x^2 - 12x + c$

4 **a** $A = \dfrac{1}{2}t^8 + c$ **b** 127.5

5 $y = \dfrac{5}{2}x^2 - 3x$

6 **a** $-\dfrac{3}{x} + c$ **b** $6\sqrt{x} + c$ **c** $\dfrac{4}{5}x^{\frac{5}{4}}$

7 **a** $x^{\frac{5}{2}}$ **b** $\dfrac{2}{7}x^{\frac{7}{2}} + c$

8 **a** $\dfrac{1}{3}x^3 + \dfrac{2}{3}x^{\frac{3}{2}} + c$ **b** $\dfrac{2}{3}x^{\frac{3}{2}} + 4x^{\frac{1}{2}} + c$

9 $y = \sqrt{x} + 2$

Exam practice 5 (page 68)

1 $-2x^{-1} + C$

2 **a** $\dfrac{dy}{dx} = 15x^{\frac{1}{2}}$ **b** $7x + 4x^{\frac{5}{2}} + C$

3 **b** $f(x) = 2x^{\frac{3}{2}} + 24x^{\frac{1}{2}} - 34$

4 **a** $y = x^3 - 10x^2 + 29x - 20$ **c** $y = x + 4$ **d** $\dfrac{14}{3}$

5 **a** $y = 5x - x^{-1} + C$ **b** $y = 12\frac{1}{2}$

6 **a** $y = 2x^{\frac{7}{2}} - x^3 + c$ **b** $y = 2x^{\frac{7}{2}} - x^3 + 4$

7 **a** $4x - 9$ **b** $2x^2 - 9x + C$

8 **a** $y = x^{\frac{3}{2}} + 4x^{-\frac{1}{2}}$

9 **a** $k = 3$ **b** $2x^{\frac{3}{2}} - x^3 + 41x + 18$ **c** $y = -x + 138$

10 **a** $y = x^3 - \frac{1}{2}x^2 - 5$

Practice exam paper (page 70)

1 $\dfrac{x^5}{5} - \dfrac{2x^3}{3} + x + C$

2 **a** $15, 13, 11$ **b** -80

3 **a** $-4 + 14\sqrt{2}$ **b** $9 - 3\sqrt{2}$

4 **a** $(2, 1)$ **b** $y - 1 = \frac{1}{2}(x - 2)$

    c $y - 1 = \frac{1}{2}(x - 2) \Rightarrow y = \frac{1}{2}x$

    Since $(0, 0)$ satisfies the equation $y = \frac{1}{2}x$, the perpendicular of AB passes through O.

5 **a** Curve goes through $(-4, 0), (-2, 0), (0, 0)$.
    b Curve goes through $(-4, 0), (0, 0), (4, 0)$.

6 **a** $x > 3.5$ **b** $0 < k < 8$

7 **a** $l + m = -2$ **b** $3l + m = 4$ **c** $m = -5$

8 **a** $x = -4, 0$

    c Eliminating y gives $(x + 2)^2 - 4 = 2x + 3 \Rightarrow x^2 + 4x + 4 - 4 - 2x - 3 = 0$
                                $\Rightarrow x^2 + 2x - 3 = 0$, as required.

    d The points of intersection are $(-3, -3)$ and $(1, 5)$.

9 **a** 65 seconds   **b** 1350 seconds
    c The number of questions in the second test is 25.

10 **a** $f(x) = x^3 - 4x^{\frac{3}{2}} - 2$
    b $f(4) = 30$
      $f'(4) = 36$
      Equation of tangent is $y - 30 = 36(x - 4) \Rightarrow y = 36x - 114$, as required.

SINGLE USER LICENCE AGREEMENT FOR CORE 1 FOR EDEXCEL CD-ROM
IMPORTANT: READ CAREFULLY

WARNING: BY OPENING THE PACKAGE YOU AGREE TO BE BOUND BY THE TERMS OF THE LICENCE AGREEMENT BELOW.

This is a legally binding agreement between You (the user or purchaser) and Pearson Education Limited. By retaining this licence, any software media or accompanying written materials or carrying out any of the permitted activities You agree to be bound by the terms of the licence agreement below.

If You do not agree to these terms then promptly return the entire publication (this licence and all software, written materials, packaging and any other components received with it) with Your sales receipt to Your supplier for a full refund.

YOU ARE PERMITTED TO:

● Use (load into temporary memory or permanent storage) a single copy of the software on only one computer at a time. If this computer is linked to a network then the software may only be used in a manner such that it is not accessible to other machines on the network.

● Transfer the software from one computer to another provided that you only use it on one computer at a time.

● Print a single copy of any PDF file from the CD-ROM for the sole use of the user.

YOU MAY NOT:

● Rent or lease the software or any part of the publication.

● Copy any part of the documentation, except where specifically indicated otherwise.

● Make copies of the software, other than for backup purposes.

● Reverse engineer, decompile or disassemble the software.

● Use the software on more than one computer at a time.

● Install the software on any networked computer in a way that could allow access to it from more than one machine on the network.

● Use the software in any way not specified above without the prior written consent of Pearson Education Limited.

● Print off multiple copies of any PDF file.

ONE COPY ONLY

This licence is for a single user copy of the software

PEARSON EDUCATION LIMITED RESERVES THE RIGHT TO TERMINATE THIS LICENCE BY WRITTEN NOTICE AND TO TAKE ACTION TO RECOVER ANY DAMAGES SUFFERED BY PEARSON EDUCATION LIMITED IF YOU BREACH ANY PROVISION OF THIS AGREEMENT.

Pearson Education Limited and/or its licensors own the software.
You only own the disk on which the software is supplied.

Pearson Education Limited warrants that the diskette or CD-ROM on which the software is supplied is free from defects in materials and workmanship under normal use for ninety (90) days from the date You receive it. This warranty is limited to You and is not transferable. Pearson Education Limited does not warrant that the functions of the software meet Your requirements or that the media is compatible with any computer system on which it is used or that the operation of the software will be unlimited or error free.

You assume responsibility for selecting the software to achieve Your intended results and for the installation of, the use of and the results obtained from the software. The entire liability of Pearson Education Limited and its suppliers and your only remedy shall be replacement free of charge of the components that do not meet this warranty.

This limited warranty is void if any damage has resulted from accident, abuse, misapplication, service or modification by someone other than Pearson Education Limited. In no event shall Pearson Education Limited or its suppliers be liable for any damages whatsoever arising out of installation of the software, even if advised of the possibility of such damages. Pearson Education Limited will not be liable for any loss or damage of any nature suffered by any party as a result of reliance upon or reproduction of or any errors in the content of the publication.

Pearson Education Limited does not limit its liability for death or personal injury caused by its negligence.

This licence agreement shall be governed by and interpreted and construed in accordance with English law.